Lecture Notes in Physics

The Lecture Notes in Physics

The series Lecture Notes in Physics (LNP), founded in 1969, reports new developments in physics research and teaching – quickly and informally, but with a high quality and the explicit aim to summarize and communicate current knowledge in an accessible way. Books published in this series are conceived as bridging material between advanced graduate textbooks and the forefront of research and to serve three purposes:

- to be a compact and modern up-to-date source of reference on a well-defined topic
- to serve as an accessible introduction to the field to postgraduate students and nonspecialist researchers from related areas
- to be a source of advanced teaching material for specialized seminars, courses and schools

Both monographs and multi-author volumes will be considered for publication. Edited volumes should, however, consist of a very limited number of contributions only. Proceedings will not be considered for LNP.

Volumes published in LNP are disseminated both in print and in electronic formats, the electronic archive being available at springerlink.com. The series content is indexed, abstracted and referenced by many abstracting and information services, bibliographic networks, subscription agencies, library networks, and consortia.

Proposals should be sent to a member of the Editorial Board, or directly to the managing editor at Springer:

Christian Caron
Springer Heidelberg
Physics Editorial Department I
Tiergartenstrasse 17
69121 Heidelberg / Germany
christian.caron@springer.com

C. Bona
C. Palenzuela-Luque
C. Bona-Casas

Elements of Numerical Relativity and Relativistic Hydrodynamics

From Einstein's Equations to Astrophysical Simulations

Second Edition

Springer

Carles Bona
Departament de Física
Universitat de les Illes Balears
Ctra Valldemossa km 7.5
E-07122 Palma de Mallorca
Spain
cbona@uib.es

Carlos Palenzuela-Luque
Max-Planck-Institut für Gravitationsphysik
(Albert Einstein Institut)
Golm
Germany

Carles Bona-Casas
Departament de Física
Universitat de les Illes Balears
Ctra Valldemossa km 7.5
E-07122 Palma de Mallorca
Spain

Bona, C. et al., *Elements of Numerical Relativity and Relativistic Hydrodynamics: From Einstein's Equations to Astrophysical Simulations*, Lect. Notes Phys. 783 (Springer, Berlin Heidelberg 2009), DOI 10.1007/978-3-642-01164-1

Lecture Notes in Physics ISSN 0075-8450
e-ISSN 1616-6361
ISBN 978-3-642-24250-2
e-ISBN 978-3-642-01164-1
DOI 10.1007/978-3-642-01164-1
Springer Dordrecht Heidelberg London New York

Softcover reprint of the hardcover 2nd edition 2009

Cover design: Integra Software Services Pvt. Ltd., Pondicherry

Printed on acid-free paper

Springer is a part of Springer Science+Business Media (www.springer.com)

To Montse, my dear wife and friend.

Para Tania, mi amor.

Preface to the Second Edition

The first edition of this book was issued by 2005, with the objective of providing basic tools for beginning graduate students interested in numerical relativity research. The size of the numerical relativity community has experienced a significant increase since then, due to the scientific breakthroughs in binary black hole simulations which started precisely by autumn 2005 with the famous Pretorius work. Perhaps this has contributed to exhaust the first printed edition in a couple of years.

This second edition provides the opportunity to include important new developments that have arisen since 2005, which we detail below. It will also be the opportunity to respond to the continuing shift in the community scientific objectives, by incorporating a new chapter on relativistic hydrodynamics and magnetohydrodynamics. We have tried to keep the focus on basic tools and formalisms, with most numerical applications being able to run in a single PC. But proper reference is also made to more advanced developments, requiring much larger computational resources.

Here is the list of the main changes and additions:

- In the first edition, there was no description of any harmonic formalism whatsoever. It was justified because this approach was not a mainstream one in 3D numerical relativity applications at that time. But things changed suddenly when Pretorius result happened to be precisely in a generalized harmonic formulation. The leading groups immediately tried to follow the same way, with diverse results. Today, both BSSN and generalized harmonic formulations are first-rank options in current binary black hole simulations. This material has been added as a new section at the end of the first chapter, dealing with the structure of the field equations. The important point of the mutual relationship between the harmonic and Z4 evolution formalisms is discussed in the third chapter, in a new section dealing with covariant formulations. Moreover, it has been possible recently to match numerical results with analytical approximations (in harmonic coordinates) for black hole simulations. This is why we include also at the end of the first chapter a concise account of approximate solution methods, which were just mentioned in the first edition.

- The Z4 formalism was used for deriving by mid-2005 (when the first edition was yet in print) some convenient damping terms for the energy–momentum constraints, together with their translation into the generalized harmonic framework. These new damping terms, which were actually incorporated in Pretorius work, are properly introduced in Chap. 3. Also, the ordering constraints arising in the passage from a second-order to a first-order (in space) system deserve a enhanced discussion in Chap. 4, in particular, the ordering constraints related with the shift derivatives, which were overlooked in the first edition. They have later shown their importance in the passage from the second-order generalized harmonic formalisms to their first-order version, with the inclusion of specific damping terms.
- We have added a new chapter dealing with non-vacuum spacetimes. It starts with the scalar field case, which has been considered as a candidate for modeling dark matter. Then we follow with sections on electromagnetic fields and on relativistic hydrodynamics. This sets the basis for the magnetohydrodynamics section, where we consider the general case, even beyond the ideal MHD one. This is a deliberate choice, as we feel that new important developments will come precisely in this area, contributing to the full explanation of many puzzling astrophysical observations.
- Concerning numerical tools, finite-volume methods should be still considered, with a view on hydrodynamical simulations. In the first edition, however, an upwind-biased variant was proposed, which required using the full eigenvector decomposition. This is not the mainstream practice today, specially in MHD applications, where the expressions for the eigenvectors get really complicated. The community is rather moving toward centered flux formulae, much more cost-efficient. In the case of the spacetime evolution, where just smooth solutions are expected, some finite-differences versions of these methods can be used with a minimal computational cost, keeping most of the robustness of the original finite-volume algorithms. Numerical methods are now included in a new specific chapter. These new tools allow for long-term black hole simulations even in normal coordinates, as described in Chap. 6.

Palma de Mallorca,
February 2009

Carles Bona
Carlos Palenzuela Luque
Carles Bona Casas

Preface to the First Edition

We got involved with numerical relativity under very different circumstances. For one of us (CB) it dates from about 1987, when the current Laser-Interferometer Gravitational Wave Observatories were just promising proposals. It was during a visit to Paris, at the Institut Henri Poincaré, where some colleagues were pushing the VIRGO proposal with such a contagious enthusiasm that I actually decided to reorient my career. The goal was to be ready, armed with a reliable numerical code, when the first detection data would arrive.

Allowing for my experience with the 3+1 formalism at that time, I started working on singularity-avoidant gauge conditions. Soon, I became interested in hyperbolic evolution formalisms. When trying to get some practical applications, I turned upon numerical algorithms (a really big step for a theoretically oriented guy) and black hole initial data. More recently, I got interested in boundary conditions and, closing the circle, again in gauge conditions. The problem is that a reliable code needs all these ingredients working fine at the same time. It is like an orchestra, where strings, woodwinds, brass, and percussion must play together in a harmonic way: a violin virtuoso, no matter how good, cannot play Vivaldi's Four Seasons by himself.

During that time, I have got many Ph.D. students. The most recent one is the other of us (CP). All of them started with some specific topic, but they needed a basic knowledge of all the remaining ones: you cannot work on the saxo part unless you know what the bass is supposed to play at the same time.

This is where this book can be of a great help. Imagine a beginning graduate student armed just with a home PC. Imagine that the objective is to build a working numerical code for simple black hole applications. The book should provide him or her with a basic insight on the most relevant aspects of numerical relativity in the first place. But this is not enough, the book should also provide reliable and compatible choices for every component: evolution system, gauge, initial and boundary conditions, even for the numerical algorithms.

This pragmatical orientation may cause this book to be seen as biased. But the idea was not to get a compendium of the excellent work that has been made on numerical relativity during these years. The idea is rather to present a well-founded and convenient way for a beginner to get into the field. He or she will quickly discover everything else.

The structure of the book reflects the peculiarities of numerical relativity research:

- It is strongly rooted in theory. Einstein's relativity is a general covariant theory. This means that we are building at the same time the solution and the coordinate system, a unique fact among physical theories. This point is stressed in the first chapter, which could be omitted by more experienced readers.
- It turns the theory upside down. General covariance implies that no specific coordinate is more special that the others, at least not a priori. But this is at odds with the way humans and computers usually model things: as functions (of space) that evolve in time. The second chapter is devoted to the evolution (or 3+1) formalism, which reconciles general relativity with our everyday perception of reality, in which time plays such a distinct role.
- It is a fertile domain, even from the theoretical point of view. The structure of Einstein's equations allows many ways of building well-posed evolution formalisms. Chapter 3 is devoted to those which are of first order in time but second order in space. Chapter 4 is devoted instead to those which are of first order both in time and in space. In both cases, suitable numerical algorithms are provided, although the most advanced ones apply mainly to the fully first-order case.
- It is challenging. The last sections of Chaps. 5 and 6 contain front-edge developments on constraint-preserving boundary conditions and gauge pathologies, respectively.[1] These are very active research topics, where new developments will soon improve the ones presented here. The prudent reader is encouraged to look for updates of these front-edge parts in the current scientific literature.

A final word. Numerical relativity is not a matter of brute force. Just a PC, not a supercomputer, is required to perform the tests and applications proposed here. Numerical relativity is instead a matter of insight. Let the wisdom be with you.

Palma de Mallorca, *Carles Bona*
January 2005 *Carlos Palenzuela Luque*

[1] Note to the second edition. The chapter numbers here correspond to the first edition. In this second edition, these tentative developments have been either removed or replaced by other material. This fact confirms the prediction we made in this first Preface.

Contents

Chapter 1
The 4D Spacetime

1.1 Spacetime geometry

Physics theories are made by building mathematical models that correspond to physical systems. General relativity, the physical theory of gravitation, models spacetime in a geometrical way: as a 4D manifold. The concept of manifold is just a generalization to the multidimensional case of the usual concept of a 2D surface. This will allow us to apply the well-known tools of differential geometry, the branch of mathematics which describes surfaces, to the study of spacetime geometry.

An extra complication comes from the fact that general relativity laws are formulated in a completely general coordinate system (this justifies the term 'General' Relativity). Special relativity, instead, makes use of inertial reference frames, where the formulation of the physical laws is greatly simplified. This means that one has to learn how to distinguish between the genuine features of spacetime geometry and the misleading effects coming from arbitrary choices of the coordinate system. This is where the curvature tensor will play a central role, as we will see in what follows.

1.1.1 The metric

We know from differential geometry that the most basic object in the spacetime geometrical description is the line element. In the case of surfaces, the line element tells us the length dl corresponding to an infinitesimal displacement between two points, which can be related by an infinitesimal change of the local coordinates x^k on the surface. In the case of the spacetime, the concept of length has to be generalized in order to include also displacements in time (which is usually taken to be the 'zero' coordinate, $x^0 \equiv ct$). This generalization is known as the 'interval' ds, which can be expressed in local coordinates as

Bona, C. et al.: *The 4D Spacetime*. Lect. Notes Phys. **783**, 1–24 (2009)
DOI 10.1007/978-3-642-01164-1_1

$$ds^2 = g_{\mu\nu} dx^\mu dx^\nu \quad (\mu, \nu = 0, 1, 2, 3)\,. \tag{1.1}$$

We can easily see from (1.1) that the tensor $g_{\mu\nu}$ is going to play a central role. In the theory of surfaces, it has been usually called 'the first fundamental form.' In general relativity it is more modestly called 'the metric' in order to emphasize its use as a tool to measure space and time intervals. The metric components can be displayed as a 4×4 matrix. This matrix is symmetric by construction (1.1), so that only 10 of the 16 coefficients are independent. Computing these 10 independent coefficients in a given spacetime domain is the goal of most numerical relativity calculations.

The metric tensor $g_{\mu\nu}$ is the basic field describing spacetime. One would need to introduce extra fields only if one wants to take into account non-gravitational interactions, like the electromagnetic or the hydrodynamical ones, but the gravitational interaction, as far as we know, can be fully described by the metric.

1.1.2 General covariance

The most interesting property of the line element (1.1) is that it is invariant under generic (smooth) changes of the spacetime coordinates, namely

$$x^\mu = F^\mu(x^{\nu'}). \tag{1.2}$$

This is because the values of space or time intervals are independent of the coordinate system one is using for labeling spacetime points. This means that the components of the metric must change in a suitable way in order to compensate the changes of the differential coefficients dx^μ in (1.1),

$$g_{\mu'\nu'} = g_{\mu\nu} \frac{\partial x^\mu}{\partial x^{\mu'}} \frac{\partial x^\nu}{\partial x^{\nu'}}. \tag{1.3}$$

We will say then that the metric transforms in a covariant way or, more briefly, that it behaves as a covariant tensor field under the general coordinate transformations (1.2).

The general covariance (1.3) of the metric means that, without altering the properties of spacetime, one can choose specific coordinate systems that enforce some interesting conditions on the metric coefficients. One can choose for instance any given (regular) spacetime point P and devise a coordinate system such that

$$g_{\mu'\nu'}\,|_{\mathrm{P}} = diag\{-c^2, +1, +1, +1\} \quad \partial_{\rho'}\, g_{\mu'\nu'}\,|_{\mathrm{P}} = 0 \tag{1.4}$$

(local inertial coordinate system at P). This means that special relativity holds true locally (in the strongest sense: a single point at a time), and it will also be of great help in shortening some proofs by removing the complication of having to deal with arbitrary coordinate systems.

At this point, we must notice some ambiguity which affects the very meaning of the term 'solution.' In the geometrical approach, one solution corresponds to one spacetime, so that metric coefficients that can be related by the covariant transformation (1.3) are supposed to describe the same metric, considered as a geometrical object, independent of the coordinate system. In this sense, we can see how in exact solutions books (see for instance [1]) different forms of the same metric appear, as discovered by different authors. In the differential equations approach, however, the term solution applies to every set of metric components that actually verifies the field equations, even if there could be some symmetry (coordinate or 'gauge') transformation relating one of these 'solutions' to another.

This is by no way a mere philosophical distinction. If general relativity has to be (as it is) general covariant, then the field equations must have two related properties:

- The equations must be unable to fully determine all the metric coefficients. Otherwise there would be no place for the four degrees of freedom corresponding to the general covariant coordinate (gauge) transformations (1.3).
- The equations must not prescribe any way of choosing the four spacetime coordinates. Otherwise there will be preferred coordinate systems and general covariance would be broken.

But in numerical relativity there is no way of getting a solution without computing the values of every metric component. This means that we must complete the differential system obtained from the field equations, by prescribing suitable coordinate conditions, before any numerical calculation can be made. The mathematical properties of the resulting complete system will of course depend on this choice of the coordinate gauge. We will come back to this point later.

1.1.3 Covariant derivatives

The very concept of derivative intrinsically involves the comparison of field values at neighboring points. The prize one has to pay for using arbitrary coordinate systems is that one can no longer compare just field components at different points: one must also compensate for the changes of the coordinate basis when going from one point to another. In this way we can interpret the two contributions that arise when computing the covariant derivative of a vector field:

$$\nabla_\mu v^\nu = \partial_\mu v^\nu + \Gamma^\nu_{\rho\mu} v^\rho . \tag{1.5}$$

The first term corresponds to the ordinary partial derivatives of the field components, whereas the second one takes into account the variation of the coordinate basis used for computing these components. The Γ symbols in (1.5) are known as 'connection coefficients' because they actually allow to compare fields at neighboring points.

The covariant derivative of tensors with 'downstairs' indices contains connection terms with the opposite sign ('downstairs' components correspond to the dual basis). In the case of the metric, for instance, one has

$$\nabla_\rho g_{\mu\nu} = \partial_\rho g_{\mu\nu} - \Gamma^\sigma_{\rho\mu} g_{\sigma\nu} - \Gamma^\sigma_{\rho\nu} g_{\mu\sigma} \tag{1.6}$$

(notice that every additional index needs its own connection term).

The connection coefficients $\Gamma^\rho_{\mu\nu}$ are not tensor fields. They transform under a general coordinate transformation (1.2) in the following way:

$$\Gamma^{\rho'}_{\mu'\nu'} = \frac{\partial x^{\rho'}}{\partial x^\rho} \left[\Gamma^\rho_{\mu\nu} \frac{\partial x^\mu}{\partial x^{\mu'}} \frac{\partial x^\nu}{\partial x^{\nu'}} + \frac{\partial^2 x^\rho}{\partial x^{\mu'} \partial x^{\nu'}} \right] . \tag{1.7}$$

The additional second derivative terms appearing in (1.7) compensate exactly the analogous terms arising in the transformation of the partial derivative contributions in (1.5) and (1.6), so that the covariant derivative of a tensor field is again a tensor field. Notice, however, that the extra second derivatives terms in (1.7) are symmetric in the lower indices. This means that the antisymmetric combinations

$$\Gamma^\rho_{[\mu\nu]} \equiv \frac{1}{2} \left(\Gamma^\rho_{\mu\nu} - \Gamma^\rho_{\nu\mu} \right) \tag{1.8}$$

correspond to the components of a tensor field (torsion tensor), because the antisymmetric part of the second derivative terms in (1.7) actually vanishes.

Coming back to the metric tensor, let us remember that one can define at any fixed point P a locally inertial coordinate system in such a way that both conditions in (1.4) hold true. It is natural to assume that the connection coefficients should also vanish in the local inertial system at P, in order to make sure that special relativity is fully recovered locally. These conditions imply that, in the local inertial coordinate system at P

- The torsion (1.8) vanishes

$$\Gamma^\rho_{[\mu\nu]} = 0 . \tag{1.9}$$

- The metric is preserved by covariant differentiation

$$\nabla_\rho g_{\mu\nu} = 0 . \tag{1.10}$$

Notice that both (1.9) and (1.10) are tensor equations. And the vanishing of any tensor quantity in a local inertial system implies that it must actually vanish in any other coordinate system. This fact, allowing for (1.6), provides a very useful expression for the connection coefficients in terms of the first derivatives of the metric components:

$$\Gamma^{\sigma}_{\mu\nu} = \frac{1}{2} g^{\sigma\rho}[\partial_{\mu} g_{\rho\nu} + \partial_{\nu} g_{\mu\rho} - \partial_{\rho} g_{\mu\nu}] \tag{1.11}$$

(Christoffel symbols), where we have noted with ‘upstairs’ indices the components of the inverse matrix of the metric, namely

$$g^{\mu\rho} g_{\rho\nu} = \delta^{\mu}_{\nu}. \tag{1.12}$$

1.1.4 Curvature

Up to this point, all we have said could perfectly apply to the special relativity (Minkowski) spacetime. All the complications with covariant derivatives and connection coefficients could arise just from using non-inertial coordinate systems. Minkowski spacetime is said to be flat because a further specialization of the local inertial coordinate system can make the metric form (1.4) to apply for all spacetime points P simultaneously.

In general relativity, in contrast, gravity is seen as the effect of spacetime curvature. So one must distinguish between the intrinsic effects of curvature (gravitation) and the sort of ‘inertia forces’ arising from weird choices of coordinate systems. Here again, this is a very well-known problem from surface theory. The curvature of a surface can be represented by its curvature tensor (Riemann tensor, as it is known in general relativity), which can be defined as follows:

$$\left(\nabla_{\rho} \nabla_{\sigma} - \nabla_{\sigma} \nabla_{\rho}\right) v^{\mu} = R^{\mu}_{\nu\rho\sigma} v^{\nu}, \tag{1.13}$$

so that it can be interpreted as a measure of the non-commutativity of (covariant) derivatives: a property that characterizes true curved spacetimes. The Riemann tensor $R^{\mu}_{\nu\rho\sigma}$ defined by (1.13) can be explicitly computed, allowing for (1.5), in terms of the connection coefficients:

$$R^{\mu}_{\nu\rho\sigma} = \partial_{\rho} \Gamma^{\mu}_{\sigma\nu} - \partial_{\sigma} \Gamma^{\mu}_{\rho\nu} + \Gamma^{\mu}_{\rho\lambda} \Gamma^{\lambda}_{\sigma\nu} - \Gamma^{\mu}_{\sigma\lambda} \Gamma^{\lambda}_{\rho\nu}. \tag{1.14}$$

It is clear from (1.14) that in a flat spacetime, where there exists a coordinate system in which all connection coefficients vanish everywhere, the curvature tensor is zero, namely

$$R^{\mu}_{\nu\rho\sigma} = 0, \tag{1.15}$$

and, like any other tensor equation, it holds in any other coordinate system. Conversely, if the tensor condition (1.15) does not hold, then (1.14) tells us that there cannot be any coordinate system in which all connection coefficients vanish everywhere and the manifold considered is not flat. It follows that (1.15) is a necessary and sufficient condition for a given spacetime to be flat. So finally we have one intrinsic and straightforward way to distinguish between genuine curved spaces and flat spaces 'disguised' in arbitrary coordinate systems.

1.1.5 Symmetries of the curvature tensor

Riemann curvature tensor is a four-index object. In 4D spacetime, this could lead up to $4^4 = 256$ components. Of course there are algebraic symmetries that contribute to reduce the number of its independent components. Part of these symmetries can be directly obtained from the generic definition (1.14), which holds for arbitrary connection coefficients. The remaining ones come from taking into account the relationship (1.11) between the connection coefficients and the metric tensor. We have summarized them in Table 1.1.

Table 1.1 Algebraic symmetries of the curvature tensor.

Generic case symmetries	Metric connection (1.11)
$R^{\mu}_{\ \nu\rho\sigma} = -R^{\mu}_{\ \nu\sigma\rho}$	$R_{\mu\nu\rho\sigma} = -R_{\nu\mu\rho\sigma}$
$R^{\mu}_{\ \nu\rho\sigma} + R^{\mu}_{\ \rho\sigma\nu} + R^{\mu}_{\ \sigma\nu\rho} = 0$	$R_{\mu\nu\rho\sigma} = R_{\rho\sigma\mu\nu}$

But, even taking all these symmetries into account, one has still 20 algebraically independent components to deal with. One can easily realize, however, that lower rank tensors can be obtained by index contraction from the Riemann tensor. Allowing for the algebraic symmetries, there is only one independent way of contracting a pair of indices of the curvature tensor, namely

$$R_{\mu\nu} \equiv R^{\lambda}_{\ \mu\lambda\nu}, \tag{1.16}$$

which is known as 'Ricci tensor' in general relativity. It follows from the algebraic properties of the Riemann tensor that (1.16) is symmetric in its two indices, so it has only 10 independent components. Contracting again in the same way, one can get the Ricci scalar

$$R \equiv R^{\lambda}_{\ \lambda} = R^{\rho\sigma}_{\ \ \rho\sigma}. \tag{1.17}$$

The Ricci tensor (1.16) and the Ricci scalar (1.17) play a major role when trying to relate curvature with the energy content of spacetime. In 3D manifolds, the Ricci tensor allows to obtain algebraically all the components of the curvature tensor (both of them have only six independent components). In the 4D case this is no longer possible: the importance of the Ricci tensor comes instead from the Bianchi identities,

$$\nabla_\lambda R^\mu_{\ \nu\rho\sigma} + \nabla_\rho R^\mu_{\ \nu\sigma\lambda} + \nabla_\sigma R^\mu_{\ \nu\lambda\rho} = 0, \tag{1.18}$$

which can be obtained directly from (1.14). One can contract two pairs of indices in (1.18) in order to get the following 'contracted Bianchi identity' for the Ricci tensor

$$\nabla_\mu \left[R^{\mu\nu} - \frac{R}{2}\, g^{\mu\nu} \right] = 0, \tag{1.19}$$

which can be interpreted as a covariant conservation law for the combination

$$G_{\mu\nu} \equiv R_{\mu\nu} - \frac{R}{2}\, g_{\mu\nu}, \tag{1.20}$$

which is known as the Einstein tensor. We will see the importance of the identity (1.19) in what follows.

1.2 General covariant field equations

General relativity is a metric theory of gravitation. This means that the physical effects of gravitation are identified with the geometrical effects of spacetime curvature. We have seen in the previous section how to describe spacetime curvature in a general covariant way, so that there are no preferred coordinate systems. In this section, we will see how to incorporate the effect of matter and non-gravitational fields. We will need first to generalize their special relativity description, which is made in terms of inertial reference frames, to general coordinate systems. Then, we will see how the energy content of these fields can be used as a source of spacetime curvature in Einstein's theory and the complexity of the resulting field equations, which motivates the use of some approximation techniques. Numerical approximations are singled out as the general purpose ones, without any underlying physical assumption which could restrict its domain of applicability.

1.2.1 The stress–energy tensor

In special relativity, the energy content of both matter and fields is described by a symmetric tensor $T^{\mu\nu}$ (stress–energy tensor). For instance, in the case of an ideal fluid, where one neglects heat transfer, viscosity, and non-isotropic pressure effects, the stress–energy tensor can be written as

$$T^{\mu\nu} = \mu\, u^\mu u^\nu + p\,(g^{\mu\nu} + u^\mu u^\nu), \tag{1.21}$$

where u^μ is the fluid four-velocity

$$u^\mu = \gamma\,(1, \boldsymbol{v}) \tag{1.22}$$

(we are using geometrized units, so that $c = 1$), and μ and p are, respectively, the energy density and the isotropic pressure of the fluid in the comoving reference frame ($\boldsymbol{v} = 0$), where the stress–energy tensor could be written as

$$T^{\mu\nu} = diag(\mu, p\,, p\,, p\,), \tag{1.23}$$

so that one can read directly the stress (isotropic pressure in this case) contribution in the space components and the energy contribution in the time component. The neglection of heat transfer implies that there cannot be momentum contributions in the comoving frame.

Energy and momentum conservation in special relativity is translated into a conservation law for $T^{\mu\nu}$, which can be written in differential form as

$$\partial_\nu T^{\mu\nu} = 0. \tag{1.24}$$

In the ideal fluid case (1.21), one can easily recover from (1.24) (the special relativistic versions of) the continuity equation and the Euler equation for ideal fluids. But (1.24) is a basic conservation law, valid in the general case, not just for the ideal fluid one. It is then natural to generalize (1.24) as

$$\nabla_\nu T^{\mu\nu} = 0, \tag{1.25}$$

so that one gets a general covariant law with the right special relativistic limit. And one is ready now to incorporate the stress–energy tensor into the general relativity framework.

1.2.2 Einstein's field equations

The general covariant conservation laws of both the Einstein and the stress–energy tensors ((1.19) and (1.25)) provide good candidates to relate curvature with the spacetime energy content. General relativity, Einstein's theory of Gravitation, is obtained when one imposes the direct relationship (Einstein's field equations):

$$G^{\mu\nu} = 8\pi\, T^{\mu\nu}, \tag{1.26}$$

where the 8π factor comes out from the Newtonian gravitation limit (we are using here geometrical units so that both the gravitational constant G and light speed c are equal to unity).

We can read (1.26) from right to left, concluding that matter or any kind of physical field acts as a gravitational source which determines the local geometry of spacetime. In this sense, solving (1.26) as the field equations will amount to determine the metric corresponding to a given matter and energy distribution.

But, conversely, we can also read (1.26) from left to right, noticing that the physical conservation law (1.25) can now be understood as mere consequence of (1.26) if one allows for the Bianchi identities (1.19). This means that the motion of matter under the action of gravitation is also governed by Einstein's equations. For instance, if we consider a dust-like test matter content, that is, an ideal fluid of incoherent (zero pressure) particles which is insensitive to any interaction other than gravitation, it follows from (1.21) and (1.25) that

$$u^\nu \nabla_\nu u^\mu = 0. \tag{1.27}$$

The equation of motion (1.27) amounts to imposing that the test particles move along the geodesic lines of spacetime geometry:

$$\frac{d^2 x^\mu}{d\lambda^2} + \Gamma^\mu_{\rho\sigma}\frac{dx^\rho}{d\lambda}\frac{dx^\sigma}{d\lambda} = 0. \tag{1.28}$$

Geodesic lines are the natural generalization of straight lines for curved spacetime (lines whose tangent vector is constant, and minimal length lines also).

If, according to Newton's first law, straight spacetime lines correspond to the inertial motion of free particles, Eq. (1.27) can be interpreted as stating that test particles in a gravitational field are also in inertial motion, but following the 'straight lines' (geodesics) of the curved spacetime geometry. Gravitation is not considered then just as one more interaction, like electromagnetism or nuclear forces, but it is identified with spacetime curvature.

Coming back to the field equations (1.26), the Bianchi identities (1.19) allow us to write

$$\nabla_0(G^{0\nu} - 8\pi\, T^{0\nu}) + \nabla_k(G^{k\nu} - 8\pi\, T^{k\nu}) = 0 \qquad (k = 1, 2, 3)\,, \tag{1.29}$$

where latin indices will refer to space coordinates. This means that the subset of four Einstein's equations with at least one time component, namely

$$G^{0\nu} = 8\pi\, T^{0\nu}, \tag{1.30}$$

are first integrals of the system. These 'energy–momentum first integrals' get preserved forever provided that the remaining six equations hold true everywhere (you can prove it first for the three space components in (1.30) and, allowing for the result, complete then the proof for the time component). This confirms that only 6 of the 10 Einstein field equations are actually independent, so that the equations do not contain enough information to determine all of the 10 independent metric coefficients, as expected from the general covariance of the theory (see Sect. 1.1.2). We will be more precise about that point in the next chapter.

1.2.3 Structure of the field equations

From now on, we will look at Einstein's equations as a set of differential equations that one must solve for the spacetime metric once the energy content of spacetime is known. It is more convenient for this purpose to write (1.26) in the equivalent form

$$R_{\mu\nu} = 8\pi \, (T_{\mu\nu} - \frac{1}{2} \, T^\lambda_{\ \lambda} \, g_{\mu\nu}), \tag{1.31}$$

that is, allowing for (1.14) and (1.16),

$$\partial_\rho \Gamma^\rho_{\mu\nu} - \partial_\mu \Gamma^\rho_{\rho\nu} + \Gamma^\rho_{\rho\lambda} \Gamma^\lambda_{\mu\nu} - \Gamma^\rho_{\lambda\mu} \Gamma^\lambda_{\rho\nu} = 8\pi \, (T_{\mu\nu} - \frac{1}{2} \, T^\lambda_{\ \lambda} \, g_{\mu\nu}), \tag{1.32}$$

where we must remember the dependence of the connection coefficients on the metric (1.11), namely

$$\Gamma^\sigma_{\mu\nu} = \frac{1}{2} \, g^{\sigma\rho} [\partial_\mu g_{\rho\nu} + \partial_\nu g_{\mu\rho} - \partial_\rho g_{\mu\nu}]. \tag{1.33}$$

It is clear that (1.32) is a non-linear system of second-order differential equations on the metric tensor. This means that generic solutions must have continuous first derivatives (metric coefficients must be smooth). This point is important when one tries to build up composite solutions, covering different regions of spacetime, by matching local solutions which hold only on one of such regions. This is a very common situation in local field theories, like electromagnetism, where different solutions are obtained for the 'interior' region, inside the charge distribution, and the 'exterior' or outside one. In the general relativity case, the matching conditions for the composite solution to be valid amount to the continuity of the metric tensor and its first partial derivatives.

A closer look at (1.32) allows one to notice that the 'principal part' (the terms containing the highest order derivatives) can be put into flux-conservative form, that is, as a four-divergence, namely

$$\partial_\rho [\Gamma^\rho_{\mu\nu} - \delta^\rho_\mu \Gamma^\sigma_{\sigma\nu}]. \tag{1.34}$$

This means that one can interpret (1.32) as a system of balance laws, like in fluid dynamics, with the principal part terms (1.34) describing transport and the remaining ones acting as sources. The right-hand-side terms, given by the stress–energy tensor, would describe sources of non-gravitational nature whereas the quadratic terms on the left-hand side

$$\Gamma^{\rho}_{\rho\lambda}\Gamma^{\lambda}_{\mu\nu} - \Gamma^{\rho}_{\lambda\mu}\Gamma^{\lambda}_{\rho\nu} \tag{1.35}$$

would describe the action of the gravitational field itself, acting as its own source.

One must be very careful, however, with this physical interpretation, because the splitting of the left-hand-side terms is not unique. Notice for instance that

$$\Gamma^{\rho}_{\rho\mu} = \frac{1}{2}\, g^{\sigma\rho}\partial_\mu g_{\sigma\rho} = \partial_\mu ln(\sqrt{g}), \tag{1.36}$$

where g stands here for the absolute value of the determinant of the metric. This allows to rearrange terms in (1.32) so that the principal part can be written as

$$\frac{1}{\sqrt{g}}\,\partial_\rho[\sqrt{g}\,(\Gamma^{\rho}_{\mu\nu} - \delta^{\rho}_{\mu}\Gamma^{\sigma}_{\sigma\nu})], \tag{1.37}$$

and the remaining quadratic terms are now

$$\Gamma^{\rho}_{\rho\mu}\Gamma^{\sigma}_{\sigma\nu} - \Gamma^{\rho}_{\lambda\mu}\Gamma^{\lambda}_{\rho\nu}, \tag{1.38}$$

instead of (1.34) and (1.35), respectively.

On the other hand, from the numerical relativity point of view, the balance law structure of (1.32) is a blessing, because one can benefit from the experience and results from a much more mature field: computational fluid dynamics (CFD). This does not mean that all CFD techniques will work fine when applied directly to numerical relativity, but at least one has a very good guidance, based on years of research. We will take advantage of this fact in our numerical simulations.

For instance, one can notice that the flux-conservative structure of the principal part of the equations allows 'weak solutions.' In the case of general relativity, this means that the metric could have first partial derivatives which are just piecewise continuous. Derivatives across the discontinuity surfaces would lead to Dirac delta terms, so that the requirement that such delta terms cancel out exactly in the field equations (1.31), when interpreted in the sense of distributions, provides the time evolution of these surfaces. It follows that the discontinuity surfaces (gravitational shock waves) must propagate with light speed.

The use we are making of the term 'shock waves' is just inspired in fluid dynamics, but is not fully justified here. This is because the principal part of our Eq. (1.32) does not contain products of the connection coefficients with their derivatives so that, in the case of weak solutions, the Dirac delta terms appear always multiplied by continuous factors. This is in contrast with the usual situation in fluid dynamics, where the principal part of the Euler equation contains convective terms of the form

$$v^k \nabla_k v^i, \tag{1.39}$$

so that discontinuities in the fluid speed $\boldsymbol{v}$ lead to Dirac delta terms with discontinuous coefficients.

Equations that, like Euler equation, contain the kind of stronger non-linearities in the principal part are said to be 'genuine non-linear' whereas equations that, like Einstein's equations, do not contain them are said to be 'linearly degenerate.' This is not a mere terminological distinction. In the genuine non-linear case, shocks can develop even from smooth initial data and their propagation speed can be either higher or lower than the characteristic speed (sound speed in fluid dynamics, where one can get either supersonic or subsonic waves). In the linearly degenerate case, in contrast, discontinuities can never arise from smooth initial data and their propagation speed is always the characteristic one (light speed in general relativity). In the fluid dynamics language, these are just 'contact discontinuities' instead of genuine shocks.

This discussion seems to suggest that Einstein's equations are in some sense easier than Euler or Navier–Stokes equations for fluid dynamics. This is true only if we look at the non-linearities of these equations from the qualitative point of view. But the situation is completely reversed if we look at it from the quantitative point of view. Remember that the basic quantities in (1.31) are not the connection coefficients, but the metric tensor. And notice that the metric derivatives in the expression (1.33) are always multiplied by the coefficients $g^{\rho\sigma}$ of the inverse matrix of the metric.

Every such coefficient is computed as the adjoint of the corresponding metric component (6 terms) divided by the metric determinant (24 terms). Every index contraction involves the 10 components of the inverse metric, that is, 60 terms (plus the 24 terms denominator). Now, two index contractions are required in the quadratic contributions

$$\Gamma^\rho_{\lambda\mu} \Gamma^\lambda_{\rho\nu} \tag{1.40}$$

in (1.35). For every fixed value of μ and ν, we can expand (1.40) in terms of first metric derivatives: five such double contractions appear. This makes $5 \times 60^2 = 18{,}000$ terms (denominators apart) per equation, that is, 1.8×10^5 terms for (the principal part of) the whole system (1.26).

A similar estimate of the remaining contributions (including second derivatives and matter terms) can raise the count up to about 2.3×10^5 in the full general case (if one multiplies everything by the square of the metric determinant in order to remove all denominators). These quarter-of-million terms provide one of the reasons why Einstein's equations deserve their reputation as possibly the hardest ones in their class.

1.3 Einstein's equations solutions

1.3.1 Symmetries: Lie derivatives

A useful strategy for simplifying the field equations system is to focus on particular solutions with some kind of symmetry. It is well known that, by adapting the coordinate system to a given symmetry of the solution, one can usually reduce the number of relevant coordinates. For instance, in the case of axial symmetry, one can take the azimuthal angle ϕ to be one of the four spacetime coordinates so that in this adapted coordinate system one has

$$\partial_\phi g_{\mu\nu} = 0, \tag{1.41}$$

and the field equations can then be written in a simpler form.

As a consequence of (1.41), all the geometrical objects that, like the curvature tensor, can be derived from the metric without further inputs must share the same symmetry, namely

$$\partial_\phi R^\mu_{\nu\rho\sigma} = 0. \tag{1.42}$$

Then, allowing for Einstein's field equations (1.31), all the physical quantities that can be computed, without further input, from the stress–energy tensor and the metric must also share the same symmetry. In the ideal fluid case (1.21), for instance, one has

$$\partial_\phi\, \tau = \partial_\phi\, p = 0\,, \qquad \partial_\phi\, u^\mu = 0, \tag{1.43}$$

so that any dependence on the azimuthal angle ϕ disappears (ϕ is an 'ignorable' coordinate).

From the group-theoretical point of view, we can identify ϕ with the parameter labeling a continuous group of transformations (rotations around one axis in this case), which is usually known as a 'Lie group.' These transformations can be interpreted as mapping every spacetime point P into a continuous set of points, one for every value of ϕ. This set of image points of a single one P defines an orbit of the group. When the mapping is continuous, this orbit is a curve on the manifold and one can compute its tangent vector field $\boldsymbol{\xi}$, which is known as the group generator. For instance, the vector field $\boldsymbol{\xi}$ that generates axial symmetry gets a trivial form in the adapted coordinate system, namely

$$\xi^\mu = \delta^\mu_{(\phi)}. \tag{1.44}$$

These group-theoretical considerations will help us in generalizing expressions like (1.41), (1.42), (1.43), and (1.44), which are valid only in the adapted coordinate system, to a general one, in keeping with the general covariance of the theory. The standard recipe of changing partial derivatives into covariant ones will not work here, because the basic equation (1.41) would be transformed in that way into an identity, valid for any spacetime independent of its symmetry properties.

The right generalization is based on the fact that the group orbits fill out spacetime: every point P is contained into its own orbit. This implies that one can just compare any tensor at P with its image under an infinitesimal transformation of the Lie group, in order to define a derivative (Lie derivative). Notice that this definition does not imply that the continuous transformations we are using should be symmetry transformations: the concepts of group orbits and generators are valid for any continuous group of transformations, not just for symmetry groups.

In the case of scalar quantities, the Lie derivative along the vector $\boldsymbol{\xi}$ reduces to the directional derivative. For instance, the first two equations in (1.43) can be written in a generic coordinate system as

$$\mathcal{L}_{\boldsymbol{\xi}}\, \tau \equiv \xi^\mu \partial_\mu\, \tau = 0\,, \qquad \mathcal{L}_{\boldsymbol{\xi}}\, p = 0. \tag{1.45}$$

In the case of vector quantities, like in the last equation in (1.43), an extra term appears, namely

$$\mathcal{L}_{\boldsymbol{\xi}}\, u^\mu \equiv \xi^\rho \partial_\rho\, u^\mu - u^\rho \partial_\rho\, \xi^\mu = 0\,. \tag{1.46}$$

Notice that one could replace partial derivatives by covariant ones in (1.46) without altering the result: this is a tensor expression, valid in a general coordinate system. The same can be done with the original equation (1.41), namely

$$\mathcal{L}_{\boldsymbol{\xi}}\, g_{\mu\nu} \equiv \xi^\rho \partial_\rho\, g_{\mu\nu} + g_{\mu\rho}\, \partial_\nu \xi^\rho + g_{\rho\nu}\, \partial_\mu \xi^\rho = 0\,, \tag{1.47}$$

where a correction term appears for every index, following the pattern of covariant derivatives, but with the opposite sign. As a consequence, expression (1.47) gets a simpler form:

$$\mathcal{L}_{\boldsymbol{\xi}}\, g_{\mu\nu} \equiv \nabla_\mu \xi_\nu + \nabla_\nu \xi_\mu = 0\,, \tag{1.48}$$

which is known as the Killing equation. Any solution $\boldsymbol{\xi}$ of the Killing equation is known as a Killing vector field and can be interpreted as the generator of a one-parameter group of isometry transformations (symmetries).

Remember that for a general coordinate transformation the metric coefficients transform in the covariant way (1.3). Isometry transformations are the particular cases such that the final coefficients happen to be identical to the original ones, revealing some symmetry of spacetime. Tensors transform in a covariant way under any change of coordinates, but they are invariant only under isometry transformations. In the case of the curvature tensor, for instance, this fact translates into the generic coordinate system version of (1.42), namely

$$\mathcal{L}_{\boldsymbol{\xi}}\, R^\mu{}_{\nu\rho\sigma} = 0\,, \tag{1.49}$$

which must hold for every Killing vector $\boldsymbol{\xi}$.

1.3.2 Exact solutions

Symmetry considerations are of great help in order to find exact solutions of Einstein's field equations (1.26). Although hundreds of particular solutions have been found (see [1] for an excellent compendium), and are still being found today, only those with a high degree of symmetry are being widely used to build astrophysical or cosmological models.

Let us consider for instance the standard cosmological models. A six-parameter symmetry group is assumed, so that the orbit of any given spacetime point P is a spatial hypersurface. The 6D symmetry group can be described as consisting of a 3D subgroup of rotations (which will leave the origin point O invariant), plus three more independent generators mapping the origin O into any other point of the same hypersurface. From the physical point of view, we can just say that spacetime is spatially homogeneous and isotropic (cosmological principle).

As this is the maximum degree of symmetry for a 3D manifold, it follows from a classical theorem that the spatial hypersurfaces must be of constant curvature. One can also align the time axis with the normal vectors to these space hypersurfaces. Putting together all these results, it follows that the line element with such maximum degree of spatial symmetry can be written as

$$-dt^2 + R(t)^2 \left(\frac{dr^2}{\sqrt{1 - k\, r^2}} + r^2\, d\Omega^2 \right) \tag{1.50}$$

(Friedman–Robertson–Walker metrics, FRW in what follows), where $R(t)$ is an arbitrary function and the parameter k can be

$$k = +1\,,\ -1\,,\ 0\,, \tag{1.51}$$

corresponding, respectively, to positive, negative, or zero curvature of the space hypersurfaces. As commented in the former subsection, all the quantities obtained from the metric without further input must share its symmetries. This means that the stress–energy tensor of the FRW metrics corresponds to an ideal fluid (1.21) with uniform energy density and pressure distribution

$$\tau = \tau(t) \quad p = p\,(t)\,. \tag{1.52}$$

The particular expressions for both the energy density τ and the pressure p will depend of course on the specific expression for the 'cosmological radius' $R(t)$ that is being used.

Another widely used solution, the Schwarzschild line element, describes a static and spherically symmetric spacetime. From the group-theoretical point of view, it can be obtained by imposing a 4D group of symmetries. One of the group generators is supposed to describe time translations, so that we can use an adapted time coordinate in which all metric components are time independent. Also, as in the previous case, the group contains a 3D subgroup of rotations around an origin point O, so that any given point P is mapped into any other point belonging to the same spherical surface with center at O.

The use of the term 'spherical surface' here is fully justified because a three-parameter subgroup is the maximum degree of symmetry for a 2D surface. These surfaces must then be of constant curvature, which is assumed to be positive in the spherical case. One can even define the Schwarzschild radial coordinate r so that the area S of such spherical surfaces is precisely

$$S = 4\pi r^2 \,. \tag{1.53}$$

The term 'area radius' can be used then for this radial coordinate, which is uniquely defined in the spherically symmetric case. The corresponding vacuum metric is then given by

$$-(1 - 2M/r)\, dt^2 + \frac{dr^2}{1 - 2M/r} + r^2\, d\Omega^2 \,, \tag{1.54}$$

where M is an arbitrary parameter (Schwarzschild mass). The Schwarzschild metric (1.54) can be used to describe spacetime in the vicinity of an isolated spherical body of mass M.

Let us remember at this point that we are talking here about local solutions. Schwarzschild spacetime, for instance, cannot be properly described as static inside the 'horizon surface' at

$$r = 2M \tag{1.55}$$

(Schwarzschild radius). The lines labeled by the t coordinate can no longer be interpreted as time lines due to the change of sign of the corresponding coefficient in (1.54). However, one can always build up a composite metric by matching a suitable interior (non-vacuum) metric to the exterior region of (1.54), outside the Schwarzschild radius. The interior metric itself need not be static: one could even use a FRW metric corresponding to pressureless fluid (dust) to model the spherical collapse of an isolated dust ball (Oppenheimer–Schneider collapse).

The same idea works backward: one can consider the Schwarzschild metric (1.54) as describing a spherical void in an expanding FRW dust universe. If the FRW metrics are homogeneous, one can get an arbitrary distribution of non-overlapping voids in this way. This is known as the Einstein–Strauss 'Swiss cheese' model, in which the static local metrics are compatible with the overall cosmological evolution.

1.3.3 Analytical and numerical approximations

Symmetry considerations can be of great help for building exact models of simple configurations. These simplified models can even serve as a guide for describing systems departing from the given symmetry by some amount: one can consider these symmetry deviations as a perturbation of the exact model. But more complex configurations, like the ones commonly encountered in astrophysics, with a lot of details to be accounted for, can be very far from any symmetry, so that perturbations around a symmetric background could not be used in a consistent way. This is why one can consider using some other approximation scheme in order to handle such more realistic models up to the required accuracy level.

The weak-field approximation scheme consists in replacing the exact line element by some perturbation series starting with the Minkowski metric of special relativity. The adimensional quotient M/R is used as the perturbation parameter, where M is the typical mass of the objects considered and R the typical distance of the configuration. This approach works fine in astrophysical scenarios involving ordinary stars, like our Sun, where

$$M = 1.47\,\mathrm{km}\,, \qquad R = 7 \times 10^5\,\mathrm{km}\,, \tag{1.56}$$

and even in the vicinity of a neutron star ($R = 10\,\mathrm{km}$ for the same mass). It can also be combined with the slow motion approximation scheme, where V/c is the adimensional parameter, V being the typical speed of the problem.

But all these schemes fail in the most extreme scenarios, where one has both strong fields and high speeds: supernova explosions, matter accretion into a black hole or the late stages of a binary system, when the two orbiting bodies merge into a single compact object. These are not just curiosities that can be left aside from our research agenda. On the contrary, these astrophysical configurations are very good candidates as gravitational wave sources. This is because the effects on Earth of gravitational waves coming from deep space are so tiny that one needs something really dramatic at the source (strong fields evolving really fast) in order to have a chance, even a small one, for detecting it. You can see [2] and [3] for an overview of the current interferometer and resonant (bar or sphere) detection facilities, respectively.

This is where numerical relativity comes into play. Numerical approximations do not rely on the smallness of physical parameters that could prevent to apply it to some otherwise interesting physical situations. The approximation here consist in a discretization process. Any function f is replaced by a finite set of values

$$f(t) \;\rightarrow\; \{f^{(n)}\} \quad (n = 1, \ldots, N)\,. \tag{1.57}$$

The term discretization comes precisely from the fact that the continuous set of values of f is replaced by a discrete (and finite) set of N numbers. The adimensional parameter related to the order of the numerical approximation is just $1/N$, independently of the physics of the problem. This is why one can apply numerical approximations to any physical situations, without having to restrict oneself to any particular dynamical regime.

The discrete set of values $\{f^{(n)}\}$ can be constructed in different ways, depending on the particular numerical approach which is being used:

- In the *spectral methods* approach, the values $f^{(n)}$ correspond to the coefficients of the development of the function f in a series of some specific set of basis functions $\{\phi^{(n)}\}$, namely

$$f = \sum_{1}^{N} f^{(n)} \phi^{(n)} . \tag{1.58}$$

- In the *finite volume* approach, the values $f^{(n)}$ correspond rather to the integrals of the function f over a set of finite domains with volume V_n, namely

$$f^{(n)} = \int_{V_n} f \, dV . \tag{1.59}$$

  Notice that it can be formally considered as a particular case of the spectral methods approach by choosing the basis functions $\phi^{(n)}$ to be zero outside the corresponding volume V_n.
- In the *finite difference* approach, the continuous spacetime itself is replaced by a lattice of points (numerical grid). The values $f^{(n)}$ are just the values of the function f for this discrete set of grid points. In the case of the time dependence, for instance, one has

$$f^{(n)}(t) = f(t_n) . \tag{1.60}$$

  This can be formally interpreted as the limit case of the finite elements approach when the volumes V_n tend to zero, so that the set of (normalized) basis functions $\{\phi^{(n)}\}$ tends to a set of Dirac delta functions.

Although all these approaches are currently used to deal with the space variables, time evolution in numerical relativity is usually dealt with finite differences (1.60). This can be interpreted as describing spacetime by a series of snapshots, step by step, like in a movie film. In the following section, we will see how to reformulate the field equations in order to keep with this description.

1.4 Harmonic formalism

Soon after Einstein's 1915 general relativity paper, the mathematical structure of the field equations (1.26) was closely investigated. The works of De Donder [4, 5] and Lanczos [6, 7] singled out a particular coordinate gauge in which this structure gets greatly simplified, taking the form of a generalized wave equation. These were later called 'harmonic coordinates' after the work of Fock [8] and were considered for a while to be the preferred coordinate system for interpreting general relativity results.

It was actually by using harmonic coordinates that the existence and uniqueness of (vacuum) Einstein's equations solutions was first proven by Choquet [9, 10]. General covariance was then invoked a posteriori in order to extend this important result to other coordinate systems.

1.4.1 The relaxed system

As we have commented in Sect. 1.2.3, the second derivative terms in Einstein's field equations (1.26) can be arranged in many different ways. Let us consider here the particular form [4–8]

$$\begin{aligned}\frac{1}{2}\, g^{\rho\sigma}\partial^2_{\rho\sigma}\, g_{\mu\nu} - \partial_{(\mu}\Gamma_{\nu)} &= -\Gamma^{\rho}_{\mu\nu}\Gamma_{\rho} \\ &+ g^{\rho\sigma}g^{\lambda\tau}[\partial_\lambda g_{\mu\rho}\,\partial_\tau g_{\nu\sigma} - \Gamma_{\mu\rho\lambda}\,\Gamma_{\nu\sigma\tau}] - 8\,\pi\,(T_{\mu\nu} - \frac{T}{2}\, g_{\mu\nu})\,,\end{aligned} \tag{1.61}$$

where indices inside round brackets are symmetrized and we have noted by Γ^μ the contracted gamma quantity, namely

$$\Gamma^\mu \equiv g^{\rho\sigma}\,\Gamma^{\mu}_{\ \rho\sigma} = -1/\sqrt{g}\,\partial_\nu(\sqrt{g}\,g^{\mu\nu}). \tag{1.62}$$

Note that we are considering here the system (1.61) for 'downstairs' metric components, although the analogous version for 'upstairs' components ($g^{\mu\nu}$) could be used as an alternative starting point [8].

As we will see below, we can now make use of coordinate freedom, by imposing the four supplementary conditions:

$$\Gamma^\mu = 0 \qquad (\mu = 0, 1, 2, 3)\,. \tag{1.63}$$

It follows that the original field equations can be written in the form

$$g^{\rho\sigma}\partial^2_{\rho\sigma}\, g_{\mu\nu} = 2\, g^{\rho\sigma}g^{\lambda\tau}[\partial_\lambda g_{\mu\rho}\,\partial_\tau g_{\nu\sigma} - \Gamma_{\mu\rho\lambda}\,\Gamma_{\nu\sigma\tau}] - 16\,\pi\,(T_{\mu\nu} - \frac{T}{2}\, g_{\mu\nu})\,, \tag{1.64}$$

so that the principal part (second derivative terms) has got the form of a generalized wave equation. In the vacuum case, we get then

$$g^{\rho\sigma}\partial^2_{\rho\sigma}\, g_{\mu\nu} = H_{\mu\nu}(g, \partial g) \tag{1.65}$$

where $H_{\mu\nu}$ is quadratic in the metric first derivatives. This fact is crucial in analytical approximation developments, as we will see below.

The simple structure (1.65) of the relaxed system allows to prove that it is well-posed [9, 10]. By a well-posed system, we mean

- that a solution exists for regular initial data
- that it is unique
- that it depends continuously on the initial data

By a solution here we mean the full set of 10 metric components. This implies that we have completely lost the coordinate gauge freedom: the relaxed system (1.64) provides one independent (second-order) evolution equation for every single metric component.

At this point, we should realize that the 'relaxed system' (1.64) is different from that of the original field equations, unless the 'harmonic constraints' condition (1.63) is additionally enforced. We use here the term constraint because (1.63) is a first-order condition that can (or not) be fulfilled by the solutions of the second-order system (1.64). Let us state this more explicitly:

- Any solution of Einstein's equations (1.61) can be written in harmonic coordinates, so that (1.63) is fulfilled. In this way we get always a solution of the relaxed system (1.64).
- The converse is not true. When the relaxed system is not general covariant, we have no extra freedom to enforce the additional condition (1.63) on generic solutions of (1.64), which cannot then be certified as true Einstein's solutions.

It follows that the solution space of the relaxed system is larger than the one containing just true Einstein's solutions in harmonic coordinates.

The hard part in the existence and uniqueness proof [9, 10] was precisely to show that the first-order harmonic constraints (1.63) are actually first integrals of the second-order relaxed system (1.64). This proved that (harmonic) Einstein's solutions constitute a well-defined subset of the solution space of (1.64): the one that can be obtained from the subset of initial data verifying the harmonic constraints (1.63). When the relaxed system is well-posed, we get in particular existence, uniqueness, and continuous dependence on the (constrained) initial data for the subspace of Einstein's solutions as a result.

1.4.2 Analytical and numerical applications

The harmonic formalism has been widely used in analytical approximation developments. This is because of the simple structure (1.65) of the relaxed system. Just as an example, let us assume for instance that the metric admits a post-Minkowskian perturbation development, namely

$$h^{\mu\nu} = \eta^{\mu\nu} + h^{\mu\nu}_{(1)} + h^{\mu\nu}_{(2)} + \cdots \,, \tag{1.66}$$

where $h^{\mu\nu}$ stands for the 'densitized' upstairs metric components, namely

$$h^{\mu\nu} \equiv \sqrt{g}\, g^{\mu\nu} \,, \tag{1.67}$$

so that, allowing for (1.62), the harmonic constraints get the simple form

$$\partial_\nu(h^{\mu\nu}) = 0\,. \tag{1.68}$$

We have considered in the previous section the $g_{\mu\nu}$ version (1.65) of the relaxed system. An equivalent $h^{\mu\nu}$ version can also be obtained [8], which is more useful in this context, namely

$$h^{\rho\sigma}\partial^2_{\rho\sigma}\, h^{\mu\nu} = M^{\mu\nu}(h, \partial h)\,. \tag{1.69}$$

We can then substitute the development (1.66) into the principal part:

$$(\eta^{\rho\sigma} + h^{\rho\sigma}_{(1)} + \cdots)\,\partial^2_{\rho\sigma}\,[\,h^{\mu\nu}_{(1)} + h^{\mu\nu}_{(2)} + \cdots] = M^{\mu\nu}(h, \partial h)\,. \tag{1.70}$$

The fact that $M^{\mu\nu}$ is quadratic in the metric derivatives allows to put (1.70) in recursive form. For any given perturbation order, we can write then

$$\Box\, h^{\mu\nu}_{(n+1)} = N^{\mu\nu}(h_{(r)}, \partial h_{(s)}) \qquad r, s \leq n\,, \tag{1.71}$$

where the box operator here stands for the flat-space d'Alembert operator, namely

$$\Box \equiv \eta^{\rho\sigma}\,\partial^2_{\rho\sigma}\,. \tag{1.72}$$

The resulting inhomogeneous wave equation (1.71) can then be integrated by standard Green function methods upto the required order.

Let us remember at this point that we started from the relaxed system (1.64), which is not equivalent to the original field equations (1.61). In the analytical perturbation framework, this translates into the fact that fulfilling the harmonic constraints (1.63) at the nth level does not imply the same thing at the next level. The fact that (1.63) is a first integral of the relaxed system, however, ensures that the integration constants we get when inverting (1.71) can be always adjusted in order to enforce the harmonic constraints at the required order, that is,

$$\partial_\nu(h^{\mu\nu}_{(n+1)}) = 0. \tag{1.73}$$

This adjustment of the harmonic constraints must be done at every order in the approximation development (1.66).

The relaxed system (1.64) is also very useful in numerical approximations. The usual practice is using explicit time-discretization algorithms. This means that the metric coefficients are computed at a given time slice, assuming that one knows their values at the previous ones. But again fulfilling the harmonic constraints (1.63) is not granted. Moreover, in numerical approximations one has no adjustable integration constants: the computer just provides specific numerical values. This means that numerical errors do make the contracted Christoffel symbols obtained from the relaxed system to depart from their assumed harmonic (zero) value, that is,

$$^{(\text{relaxed})}\Gamma^\mu \neq 0\,. \tag{1.74}$$

One can use then the non-zero values (1.74) for monitoring the quality of the numerical simulation. We will discuss further this 'free evolution' (unconstrained) approach in Chap. 3.

1.4.3 Harmonic coordinates

The term 'harmonic coordinates' comes from the way in which the harmonic constraints (1.63) can be related with the coordinate gauge. Let us consider a set of four independent harmonic functions $f_{(a)}$, that is,

$$\Box\, f_{(a)} = 0 \qquad (a = 0, 1, 2, 3)\,, \tag{1.75}$$

where the box stands now for the general relativistic wave operator, namely

$$\Box\, f \equiv g^{\mu\nu}\nabla_\mu\nabla_\nu\, f = 1/\sqrt{g}\; \partial_\mu\,[\,\sqrt{g}\, g^{\mu\nu}\partial_\nu\, f\,]\,. \tag{1.76}$$

The solutions of (1.75) can be expressed of course in terms of the chosen (generic) spacetime coordinate system. Let us now use these four harmonic functions $f_{(a)}$ to set up a new coordinate system, by taking

$$x^\mu \equiv \delta^\mu_{(a)} f_{(a)}\,. \tag{1.77}$$

In this 'harmonic' coordinate system we will have, allowing for (1.75),

$$1/\sqrt{g}\; \partial_\mu\,[\,\sqrt{g}\, g^{\mu\nu}\,] = -\Gamma^\nu = 0\,, \tag{1.78}$$

and the constraints (1.63) are verified as a result.

Note also that (1.63) is not a vector equation, as it follows from the transformation properties (1.7). Apart from linear transformations, there is not much coordinate freedom left in harmonic coordinates. This can be a problem in some cases. In order to illustrate this point, we will consider for instance a generic spherically symmetric line element, namely

$$ds^2 = -\alpha^2\, dt^2 + X^2\, dr^2 + Y^2\, \left[d\theta^2 + sin^2(\theta)\, d\varphi^2\right]\,, \tag{1.79}$$

where all the metric functions (α, X, Y) depend only on (t, r). In this case, one gets

$$\Gamma^\varphi = 0\,, \qquad \Gamma^\theta = -\frac{cot(\theta)}{Y^2} \neq 0\,, \tag{1.80}$$

so that spherical coordinates happen to be incompatible with the harmonicity (1.75).

This simple example shows that more coordinate freedom would be welcome, provided that we do not lose the benefits of the harmonic formulation. Let us perform a general coordinate transformation, starting from a harmonic coordinates system. Allowing for (1.7), we will get after some algebra

$$\Gamma^{\rho'} = g^{\mu'\nu'} [\frac{\partial x^{\rho'}}{\partial x^{\rho}} \frac{\partial^2 x^{\rho}}{\partial x^{\mu'} \partial x^{\nu'}}] = -g^{\mu\nu} [\frac{\partial^2 x^{\rho'}}{\partial x^{\mu} \partial x^{\nu}}] = -\Box\, x^{\rho'} . \tag{1.81}$$

By replacing this expression for Γ' into Einstein's field equations (1.61), we get the 'generalized harmonic system' [11]

$$\frac{1}{2}\, g^{\rho\sigma} \partial^2_{\rho\sigma}\, g_{\mu\nu} + \partial_{(\mu} H_{\nu)} - \Gamma^{\rho}_{\mu\nu} H_{\rho} \tag{1.82}$$
$$= g^{\rho\sigma} g^{\lambda\tau} [\partial_{\lambda} g_{\mu\rho}\, \partial_{\tau} g_{\nu\sigma} - \Gamma_{\mu\rho\lambda}\, \Gamma_{\nu\sigma\tau}] - 8\, \pi\, (T_{\mu\nu} - \frac{T}{2}\, g_{\mu\nu}) ,$$

where we have dropped the primes, and the 'gauge sources' H^{μ} are defined as

$$H^{\mu}(x, g) \equiv -\Gamma^{\mu} , \tag{1.83}$$

with Γ^{μ} given by (1.81). The term 'gauge sources' comes from the fact that the coordinate functions are defined now by the inhomogeneous wave equation

$$\Box\, f_{(a)} = H_{(a)} \qquad (a = 0, 1, 2, 3) , \tag{1.84}$$

instead of (1.75).

Allowing for (1.80), the generalized harmonic condition (1.84) is good enough to cover our spherical symmetry example (1.79). Also, if the gauge sources can be prescribed without involving metric derivatives, the extra terms in (1.82) do not belong to the principal part. One can then show that the generalized system is well-posed [11] in the same way as before.

It is currently stated that the generalized harmonic system (1.82) is able to deal with generic gauge choices. This is true in the following sense:

- Any coordinate system can be locally related with a harmonic system. The explicit expression of these harmonic coordinates can be actually obtained by solving the general relativistic wave equations (1.75).
- Knowing the explicit transformation, one can always compute the corresponding gauge sources from (1.81).

But in practical applications coordinate systems are not usually defined by their relationship with harmonic coordinates. Moreover, in numerical simulations one is usually trying to compute metrics that are not known in advance. One would like to choose a priori coordinate conditions in a way that could ensure some desired property of the solution. This is hard to do by dealing with gauge sources. We will see other ways to do this in the next chapter.

References

1. H. Stephani et al. *Exact Solutions of Einstein's Field Equations*, 2nd edn, Cambridge, Cambridge New York (2003)
2. http://www.ligo.caltech.edu
   http://www.virgo.infn.it
   http://geo600.aei.mpg.de
   http://tamago.mtk.nao.ac.jp
   http://www.gravity.uwa.edu.au
3. http://www.minigrail.nl/
   http://www.roma1.infn.it/rog/explorer/explorer.html
   http://gravity.phys.lsu.edu/
   http://www.gravity.uwa.edu.au
   http://www.das.inpe.br/ graviton
4. T. De Donder, *La Gravifique Einstenienne*, Gauthier-Villars, Paris (1921).
5. T. De Donder, *The Mathematical Theory of Relativity*, Massachusetts Institute of Technology, Cambridge (1927).
6. K. Lanczos, Ann. Phys. **13**, 621 (1922).
7. K. Lanczos, Z. Phys. **23**, 537 (1923).
8. V. A. Fock, *The Theory of Space, Time and Gravitation*, Pergamon, London (1959).
9. Y. Choquet (Fourès)-Bruhat, Acta Matematica **88**, 141 (1952).
10. Y. Choquet (Fourès)-Bruhat, 'Cauchy problem' in *Gravitation: An Introduction to Current Research*, ed. by L. Witten, Wiley, New York (1962).
11. H. Friedrich, Commun. Math. Phys. **100**, 525 (1985).

Chapter 2
The Evolution Formalism

2.1 Space plus time decomposition

The general covariant approach to general relativity is not adapted to our experience from everyday life. The most intuitive concept is not that of spacetime geometry, but rather that of a time succession of space geometries. This 'flowing geometries' picture could be easily put into the computer, by discretizing the time coordinate, in the same way that the continuous time flow of the real life is coded in terms of a discrete set of photograms in a movie.

In this sense, we can say that general relativity theory, when compared with other physical theories like electromagnetism, has been built upside down. In Maxwell theory one starts with the everyday concepts of electric charges, currents, electric and magnetic fields. One can then write down a (quite involved) set of field equations, Maxwell equations, that can be easily interpreted by any observer. Only later some 'hidden symmetry' (Lorentz invariance) of the solution space is recognized, and this allows to rewrite Maxwell equations in a Lorentz-covariant form. But the price to pay is gluing charges and currents on one side, and electric and magnetic fields on the other, into new 4D objects that obscure the direct relation to experience the original (3D) pieces.

In general relativity, we have started from the top, at the 4D level, so we must go downhill, in the opposite sense:

- by selecting a specific (but generic) time coordinate;
- by decomposing every 4D object (metric, Ricci, and stress–energy tensors) into more intuitive 3D pieces;
- by writing down the (much more complicated) field equations that translate the manifestly covariant ones (1.26) in terms of these 3D pieces.

General covariance will then become a hidden feature of the resulting '3+1 equations.' The equations themselves will no longer be covariant under a general coordinate transformation. But, as the solution space is not being modified by the decomposition, general coordinate transformations will still

Bona, C. et al.: *The Evolution Formalism*. Lect. Notes Phys. **783**, 25–48 (2009)
DOI 10.1007/978-3-642-01164-1_2

map solutions into solutions (as it happens with Lorentz transformations in Maxwell equations). The underlying invariance of the equations under general coordinate transformations is then preserved when performing the 3+1 decomposition. General covariant 4D equations just show up this invariance in an explicit way.

2.1.1 A prelude: Maxwell equations

Maxwell equations are usually written as

$$\nabla \cdot \boldsymbol{E} = 4\pi q, \tag{2.1}$$

$$\nabla \cdot \boldsymbol{B} = 0, \tag{2.2}$$

$$-\partial_t \boldsymbol{E} + \nabla \times \boldsymbol{B} = 4\pi \boldsymbol{J}, \tag{2.3}$$

$$\partial_t \boldsymbol{B} + \nabla \times \boldsymbol{E} = 0, \tag{2.4}$$

where it is clear that the charge and current densities, q and $\boldsymbol{J}$, act as the sources of the electric and magnetic fields, $\boldsymbol{E}$ and $\boldsymbol{B}$. We assume here for simplicity the vanishing of both the electric and magnetic susceptibilities, so that $\boldsymbol{D} = \boldsymbol{E}$ and $\boldsymbol{B} = \boldsymbol{H}$.

The second pair of equations (2.3) and (2.4) can be interpreted as providing a complete set of time evolution equations for the electric and magnetic fields (evolution system), whereas the first two equations (2.1) and (2.2) do not contain time derivatives and can be interpreted then as constraints. A straightforward calculation shows that these constraints are first integrals of the evolution system, as a consequence of the charge continuity equation:

$$\partial_t q + \nabla \cdot \boldsymbol{J} = 0\,. \tag{2.5}$$

Now, we can start joining pieces. The charge and current densities can be combined to form a four-vector I^μ,

$$I^\mu \equiv (q,\, \boldsymbol{J})\,. \tag{2.6}$$

The electric and magnetic fields can be combined in turn to form an antisymmetric tensor, namely

$$F_{\mu\nu} \equiv \begin{pmatrix} 0 & -E_j \\ E_i & F_{ij} \end{pmatrix}, \qquad F_{ij} \equiv \epsilon_{ijk} B^k \tag{2.7}$$

(electromagnetic field tensor).

The pair (2.1) and (2.3) of Maxwell equations can then be written in the manifestly covariant form

$$\nabla_\nu\, F^{\mu\nu} = 4\pi\, I^\mu\,, \tag{2.8}$$

whereas the other pair (2.2) and (2.4) can be written as

$$\nabla_\rho F_{\mu\nu} + \nabla_\mu F_{\nu\rho} + \nabla_\nu F_{\rho\mu} = 0 . \tag{2.9}$$

Note that, allowing for the antisymmetry of the electromagnetic tensor, the four-divergence of (2.8) leads immediately to the covariant version of the charge continuity equation (2.5), namely

$$\nabla_\mu I^\mu = 0 . \tag{2.10}$$

This strongly reminds the general relativity situation, where Einstein's equations (1.26), allowing for the Bianchi identities (1.19), lead to the conservation of the stress–energy tensor.

2.1.2 Spacetime synchronization

Coming back to the general relativity case, the 3+1 spacetime decomposition is based on two main geometrical elements:

- The first one is the choice of a **synchronization**. This amounts to foliate spacetime by a family of spacelike hypersurfaces, so that any spacetime

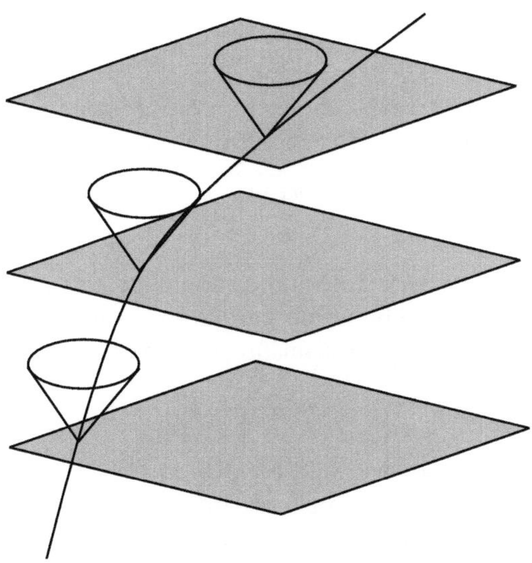

Fig. 2.1 The *time slicing* and one possible *time lines*, with some *light cones* drawn in order to show the causal character. The specific *time line* shown here starts being timelike, then changes to spacelike. The only requirement on *time lines* is that they cannot be tangent to the spacelike *slices:* they must 'thread' them. Note that, in this way, the slicing provides a natural choice of the affine parameter along time lines.

point belongs to one (and only one) slice. This geometrical construction can be easily achieved by selecting a regular spacetime function

$$\phi(x^\mu), \tag{2.11}$$

such that the level hypersurfaces of ϕ (defined by $\phi = \text{const}$) provide the desired foliation. In order to consider ϕ as a local time coordinate, we must make sure that the resulting slicing is spacelike, namely

$$g^{\mu\nu}\,\partial_\mu\phi\,\partial_\nu\phi < 0\,. \tag{2.12}$$

Every single slice can be considered as a 3D manifold. It is clear that curves on this manifold are also spacetime curves and therefore the spacetime metric can be directly used for measuring lengths on any 3D slice. In that way, the 3D metric γ_{ij} on every slice is induced by the spacetime metric $g_{\mu\nu}$. The overall picture can be easily understood by fully identifying the function ϕ with our time coordinate, that is,

$$\phi(x^\mu) \equiv t\,, \qquad \partial_\mu\phi = \delta^{(0)}_\mu\,, \tag{2.13}$$

so that the 3D line element

$$dl^2 = \gamma_{ij}\;dx^i dx^k \qquad (i,\,j = 1,2,3) \tag{2.14}$$

can be obtained from the 4D one by restricting it to the constant time surfaces, namely

$$\gamma_{ij} = g_{ij}\,. \tag{2.15}$$

- The second ingredient is the choice of a **congruence of time lines**. The simplest way is to get it as the integral curves of the system

$$\frac{dx^\mu}{d\lambda} = \xi^\mu\,, \tag{2.16}$$

so that the congruence is fully determined by the choice of the field of its tangent vectors ξ^μ. The affine parameter λ in (2.16) can be chosen to match the spacetime synchronization by imposing

$$\xi^\mu\partial_\mu\phi = 1\,. \tag{2.17}$$

Note that (2.17) is a very mild condition. It just requires that the time lines are not tangent to the constant time slices. It does not even demand the time lines to be timelike, in contrast with the stronger requirement (2.12) for the time slices (see Fig. 2.1).

Again, the meaning of (2.17) is more transparent if we use ϕ as the local time coordinate, that is (2.13),

$$\xi^\mu = \delta^\mu_{(0)}\,, \tag{2.18}$$

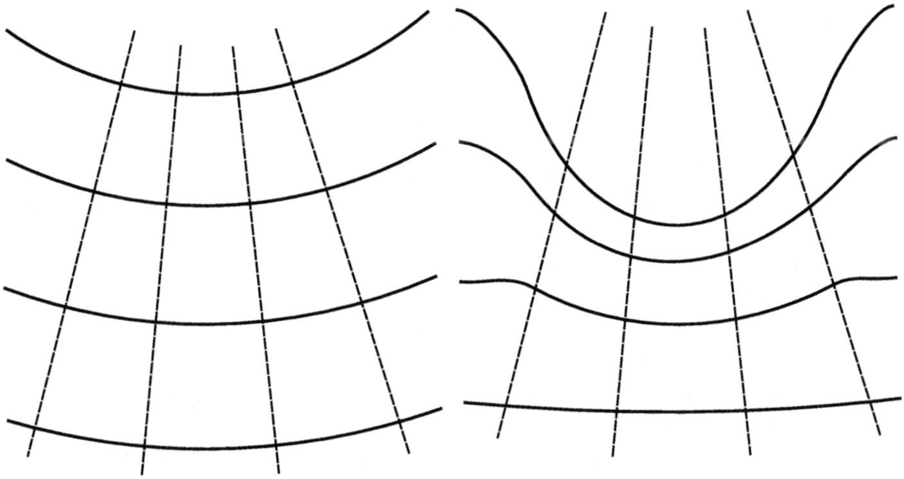

Fig. 2.2 Two different time foliations of the same spacetime. We see on the *left* a geodesic slicing. Coordinate time (represented by the *continuous level lines*) flows here homogeneously with the proper time along *normal (dashed) lines*, meaning that the lapse function is constant in this case. A coordinate singularity, the focusing of *normal lines*, is going to form in a finite coordinate time. We show in the *right side* a lapse function α such that proper time evolution slows down in the central region, while keeping a more uniform rate elsewhere. The *dashed lines* represent just the same freely falling observers, which are no longer normal to the constant time slices. This lapse-related degree of freedom will be of great help in numerical simulations of gravitational collapse, where we will like to slow down the proper time rate in the regions where a singularity is going to form (singularity avoidance).

so that the time lines equation (2.16) can be trivially integrated

$$x^0 = t \qquad x^i = \text{constant}. \tag{2.19}$$

We will focus first on the analysis of the time slicing. To do this, we will use normal coordinates, so that the time lines are chosen to be normal to the constant time slices. In this way we refrain from using the three extra degrees of freedom that would allow us to freely choose the tangent vector field ξ. Here we will not use any other ingredient than the slicing itself: we are restricting ourselves in this way to the simplest choice

$$\xi_\mu \sim \partial_\mu \phi\,, \tag{2.20}$$

or, in local adapted coordinates (2.13),

$$g_{0i} = 0 \tag{2.21}$$

(normal coordinates). The line element can then be written, allowing for (2.15) and (2.21), as

$$ds^2 = -\alpha^2 dt^2 + \gamma_{ij}\, dx^i dx^j\,, \tag{2.22}$$

where the factor α relates the coordinate time t with the proper time τ along the (normal) time lines:

$$d\tau = \alpha\, dt\,. \tag{2.23}$$

This means that the factor α (lapse function) gives us the rate at which proper time is elapsed along the normal lines (the time lines in normal coordinates).

Note that the lapse function can take different values at different spacetime points. This means that the amount of proper time elapsed when going from one slice to another can depend on the location. On the contrary, the amount of elapsed coordinate time is, by construction, independent of the space location. The particular case in which the lapse function α is constant corresponds to the geodesic slicing (see Fig. 2.2); the name will be justified in the next subsection. The combination of geodesic slicing plus normal space coordinates is known as the Gauss coordinate system.

2.1.3 The Eulerian observers

As stated before, in normal coordinates the congruence of time lines is provided by the slicing itself. We can view this congruence as the world lines of a field of observers which are at rest with respect to the spacetime synchronization (Eulerian observers). Their four-velocity field u^μ coincides, up to a sign, with the field of unit normals n^μ to the constant time slices

$$n_\mu = \alpha\, \partial_\mu \phi \qquad g^{\mu\nu} n_\mu n_\nu = -1\,. \tag{2.24}$$

The relative sign comes from the normalization condition (2.17), that is,

$$u^\mu = -n^\mu \qquad u^\mu n_\mu = 1\,. \tag{2.25}$$

In normal adapted coordinates, we have

$$u^\mu = \frac{1}{\alpha}\, \delta^\mu_{(0)}\,, \qquad n_\mu = \alpha\, \delta^{(0)}_\mu\,, \tag{2.26}$$

so that the tangent vector field u points forward in time.

The motion of any set of observers, represented by a congruence of time lines, can be decomposed into different kinematical pieces, namely

$$\nabla_\mu u_\nu = -u_\mu \dot{u}_\nu + \omega_{\mu\nu} + \chi_{\mu\nu}\,, \tag{2.27}$$

where every piece describes a different feature of the motion:

- **Acceleration**, described by the four-vector

$$\dot{u}_\mu \equiv u^\rho \nabla_\rho u_\mu\,. \tag{2.28}$$

It is the only non-trivial projection of (2.27) along u^μ.

- **Vorticity**, described by the antisymmetric tensor $\omega_{\mu\nu}$. It is the antisymmetric part of the projection of (2.27) orthogonal to u^μ ($\omega_{\mu\nu} u^\nu = 0$).
- **Deformation**, described by the symmetric tensor $\chi_{\mu\nu}$. It is the symmetric part of the projection of (2.27) orthogonal to u^μ ($\chi_{\mu\nu} u^\nu = 0$). It can be further decomposed into its trace, the expansion scalar

$$\theta \equiv tr(\chi)\,, \tag{2.29}$$

and its traceless part, the shear tensor

$$\sigma_{\mu\nu} = \chi_{\mu\nu} - \frac{\theta}{3}\,(g_{\mu\nu} + u_\mu u_\nu)\,. \tag{2.30}$$

In the case of the Eulerian observers, there is no vorticity because, by construction, they are orthogonal to the constant time hypersurfaces. Their motion is then characterized only by the acceleration vector and the deformation tensor. In adapted normal coordinates, allowing for (2.26), the acceleration vector is given by

$$\dot{u}_\mu = (0,\, \partial_i\, ln\alpha)\,, \tag{2.31}$$

so that the choice of a constant lapse corresponds to the inertial motion (free fall) of the Eulerian observers (this justifies the term 'geodesic slicing' we used in the previous subsection for the $\alpha =$ constant case).

The deformation tensor of the Eulerian observers consists also of space components only when written in adapted normal coordinates, namely,

$$\chi_{\mu\nu} = \begin{pmatrix} 0 & 0 \\ 0 & -K_{ij} \end{pmatrix}\,. \tag{2.32}$$

The 3D symmetric tensor K_{ij} in (2.32) is known as the extrinsic curvature of the slicing, whereas the minus sign in (2.32) arises from the sign convention (2.25).

The extrinsic curvature can be easily computed from (2.27). In normal adapted coordinates (2.26), we have

$$K_{ij} = -\frac{1}{2\alpha}\,\partial_t\,\gamma_{ij}\,. \tag{2.33}$$

Notice that K_{ij} admits then a double interpretation:

- From the time lines point of view, it provides the deformation $\chi_{\mu\nu}$ of the congruence of normal lines, as it follows from (2.27) and (2.32).
- From the slices point of view, it provides, up to a one half-factor, the Lie derivative of the induced metric γ_{ij} along the field of unit normals n^μ, as it follows from (2.26) and (2.33), and the space components of the 4D identity

$$\mathcal{L}_n(g_{\mu\nu}) = \nabla_\mu n_\nu + \nabla_\nu n_\mu\,. \tag{2.34}$$

Of course, these two points of view are equivalent, because the congruence of normal lines can be obtained from the slicing in a one-to-one way.

2.2 Einstein's equations decomposition

2.2.1 The 3+1 form of the field equations

Let us summarize the results of the previous section:

- We have decomposed the 4D line element into the 3+1 normal form (2.22), where the distinct geometrical meaning of the lapse function α and the induced metric γ_{ij} has been pointed out. This is analogous to decompose the electromagnetic tensor into its electric and magnetic field components.
- Einstein's field equations, contrary to Maxwell ones, are of second order. This means that one needs also to decompose the first derivatives of the 4D metric. We have started doing so in the previous subsection, where we have identified the pieces describing either the acceleration or the deformation tensor of the Eulerian observers (the lapse gradient and the extrinsic curvature K_{ij} of the slices). The remaining first derivatives can be easily computed in terms of the pieces we have got (see Table 2.1, where the full set of connection coefficients is displayed).

Table 2.1 The 3+1 decomposition of the 4D connection coefficients. Notice that the symbol $\widehat{\Gamma}^{\mu}_{\rho\sigma}$ stands for the connection coefficients of the 4D metric, whereas in what follows we will note as Γ^k_{ij} the connection coefficients of the induced 3D metric γ_{ij}.

$\widehat{\Gamma}^0_{00} = \partial_t \, ln(\alpha)$	$\widehat{\Gamma}^k_{00} = \alpha\gamma^{kj}\partial_j \, \alpha$
$\widehat{\Gamma}^0_{i0} = \partial_i \, ln(\alpha)$	$\widehat{\Gamma}^k_{i0} = -\alpha\gamma^{kj} K_{ij}$
$\widehat{\Gamma}^0_{ij} = -1/\alpha \, K_{ij}$	$\widehat{\Gamma}^k_{ij} = \Gamma^k_{ij}$

We have then for the moment a complete decomposition of the 'left-hand side' of the field equations. The corresponding decomposition of the source terms is just the well-known decomposition of the 4D stress–energy tensor $T_{\mu\nu}$ into parts which are either longitudinal (aligned with n_μ), transverse (orthogonal to n_μ), or of a mixed type, namely

- The **energy density**

$$\tau \equiv T^{\mu\nu} n_\mu n_\nu \tag{2.35}$$

- The **momentum density**

$$S_i \equiv T^{\mu}_{\ i} n_\mu \tag{2.36}$$

- The **stress tensor**

$$S_{ij} \equiv T_{ij}\,, \tag{2.37}$$

whose names arise from the physical interpretation that can be made from the point of view of the Eulerian observers.

Now we are in a position to translate the 4D field equations (1.26) in terms of the 3+1 quantities. We will reproduce here for clarity the equivalent (1.31) equations in terms of the 4D connection coefficients, so that we can apply the results of Table 2.1 in a straightforward way:

$$\partial_\rho \widehat{\Gamma}^\rho{}_{\mu\nu} - \partial_\mu \widehat{\Gamma}^\rho{}_{\rho\nu} + \widehat{\Gamma}^\rho{}_{\rho\lambda} \widehat{\Gamma}^\lambda{}_{\mu\nu} - \widehat{\Gamma}^\rho{}_{\lambda\mu} \widehat{\Gamma}^\lambda{}_{\rho\nu} = 8\pi\,(T_{\mu\nu} - \frac{1}{2} T^\lambda{}_\lambda\, g_{\mu\nu}). \tag{2.38}$$

The space components of (2.38) can then be written, after some algebra, as

$$\frac{1}{\alpha}\,\partial_t K_{ij} = -\frac{1}{\alpha}\,\nabla_i \alpha_j + R_{ij} - 2K^2_{ij} + trK\,K_{ij} - 8\pi\,[\,S_{ij} - \frac{1}{2}\,(trS - \tau)\,\gamma_{ij}\,]\,, \tag{2.39}$$

where the covariant derivatives and the Ricci tensor on the right-hand side are the ones obtained by considering every slice as a single 3D surface with metric γ_{ij} (traces are taken with the inverse matrix γ^{ij}). The same can be done with the mixed $(0i)$ components, namely

$$0 = \nabla_j\,(K_i{}^j - trK\,\delta_i{}^j) - 8\pi\,S_i\,, \tag{2.40}$$

where we get a first surprise: no time derivative appears on the left-hand side.

The remaining component of (2.38), the (00) one, leads in turn to

$$\frac{1}{\alpha}\,\partial_t\,trK = -\frac{1}{\alpha}\,\triangle\alpha + tr(K^2) + 4\pi\,(trS + \tau)\,. \tag{2.41}$$

This is also surprising, because the time derivative of the trace of K_{ij} can be obtained also from the space components equation (2.39). If we do so, we get by substituting the result into (2.41),

$$0 = tr\;R + (tr\;K)^2 - tr(K^2) - 16\pi\,\tau\,, \tag{2.42}$$

where again no time derivative appears, like in (2.40).

2.2.2 3+1 Covariance

General covariance is lost when decomposing the 4D field equations (2.38) into their 3+1 pieces ((2.39), (2.40), and (2.42)). If the solution space has not been changed in the process, general coordinate transformations still map solutions into solutions: the underlying invariance of the theory is intact. The

4D, general covariant version (2.38) just makes this underlying invariance manifest.

This does not mean, however, that covariance is completely lost. A closer look to the right-hand-side terms of the 3+1 system (2.39), (2.40), (2.41), and (2.42) shows that they are actually covariant under general space coordinate transformations

$$y^i = F^i(x^j, t)\,, \tag{2.43}$$

which preserve the geometry of every single slice. They are also unchanged under an arbitrary time coordinate rescaling

$$t' = G(t)\,, \tag{2.44}$$

which affects just the labeling of the slices, but not the slicing itself. We will call in what follows '3+1 covariance' the covariance under the restricted set of slicing-preserving coordinate transformations (2.43) and (2.44).

The 3+1 covariance of the right-hand sides of the system (2.39), (2.40), (2.41), and (2.42) follows from the fact that they are composed of two kinds of geometrical objects:

- 3D tensors, like the metric γ_{ij} or the Ricci tensor R_{ij}, which are intrinsically defined by the geometry of the slices, when considered as individual manifolds;
- pieces which can be obtained from 4D tensors by using the field of unit normals n_μ, which is intrinsically given by the slicing. This is the case of the three-acceleration $\partial_i\, ln\alpha$, the deformation (extrinsic curvature) K_{ij}, and the different projections of the stress–energy tensor.

This means that, in spite of the fact that we have used normal coordinates in their derivation, Eqs. (2.40) and (2.42) keep true in a generic coordinate system. Before getting a similar conclusion about the tensor equation (2.39), which contains a time derivative, let us consider the case of the simpler scalar equation (2.41). We know from the previous arguments that the right-hand side term will behave as a 3+1 scalar. We will consider now the transformation properties of the time derivative in the left-hand side step by step:

- It transforms under (2.43) as

$$\left(\frac{\partial\, trK}{\partial\, t}\right)_{x=\mathrm{const}} = \left(\frac{\partial\, trK}{\partial\, t}\right)_{y=\mathrm{const}} - \beta^k \left(\frac{\partial\, trK}{\partial\, y^k}\right)_{t=\mathrm{const}}, \tag{2.45}$$

  where we have introduced the shift β^k:

$$\beta^k(y,t) \equiv \left(\frac{\partial\, y^k}{\partial\, x^r}\right)\left(\frac{\partial\, x^r}{\partial\, t}\right)_{y=\mathrm{const}}. \tag{2.46}$$

- Concerning the time rescaling (2.44), let us notice that the lapse function is not a 3+1 scalar. It follows from its very definition (2.22) that it will transform instead as

$$\alpha' = \alpha \left(\frac{\partial t}{\partial t'} \right) , \tag{2.47}$$

so that the combination

$$\frac{1}{\alpha}\, \partial_t \tag{2.48}$$

is preserved. Note that the rescaling factor in (2.47) is independent of the space coordinates, so that the three-acceleration $\partial_i\, ln\alpha$ transforms as a 3+1 vector, as expected.

Putting these results together, it follows that the generic coordinate form of Eq. (2.41) can be obtained from their expression in normal coordinates by the following replacement:

$$\frac{1}{\alpha}\, \partial_t\, trK \quad \rightarrow \quad \frac{1}{\alpha}\, (\partial_t - \beta^k \partial_k)\; trK\,. \tag{2.49}$$

The 3+1 covariance of the resulting expression is clear if we notice that it has the intrinsic meaning of 'taking the proper time derivative of trK along the normal lines,' no matter what is our coordinate system. The same idea can lead to the corresponding generalization of the tensor equation (2.39), or any other in 3+1 form, by using as a rule of thumb the generic replacement

$$\frac{1}{\alpha}\, \partial_t \quad \rightarrow \quad \frac{1}{\alpha}\, (\partial_t - \mathcal{L}_\beta)\,. \tag{2.50}$$

2.2.3 Generic space coordinates

It follows from the previous considerations that the full set of Einstein's field equations can be decomposed in a generic coordinate system as follows:

$$\frac{1}{\alpha}\, (\partial_t - \mathcal{L}_\beta)\; \gamma_{ij} = -2K_{ij} \tag{2.51}$$

$$\frac{1}{\alpha}\, (\partial_t - \mathcal{L}_\beta)\; K_{ij} = -\frac{1}{\alpha}\, \nabla_i\, \alpha_j + R_{ij} - 2K^2_{ij} + trK\; K_{ij} - 8\pi \left[\, S_{ij} - \frac{1}{2}\, (trS - \tau)\; \gamma_{ij} \,\right] \tag{2.52}$$

$$0 = \nabla_j\, (K_i{}^j - trK\; \delta_i{}^j) - 8\pi\, S_i \tag{2.53}$$

$$0 = tr\, R + (tr\, K)^2 - tr(K^2) - 16\pi\, \tau\,. \tag{2.54}$$

A simpler version, in Gauss coordinates ($\alpha = 1$, $\beta = 0$), was obtained by Darmois [1]. The normal coordinates version (2.39), (2.40), (2.41), and (2.42) is due to Lichnerowicz [2]. It was extended to the general case, although in the tetrad formalism, by Choquet-Bruhat [3]. See [4] for an interesting historical review.

Table 2.2 Same as Table 2.1 for the generic coordinates case. The symbol ∇ stands here for the covariant derivative with respect to the induced metric γ_{ij}.

$\widehat{\Gamma}^0_{00} = (\partial_t \alpha + \beta^k \alpha_k - K_{ij}\beta^i\beta^j)/\alpha$	$\widehat{\Gamma}^0_{i0} = (\partial_i\, \alpha - K_{ij}\beta^j)/\alpha$
$\widehat{\Gamma}^k_{00} = \gamma^{kj}[\partial_t\beta_j + \alpha\,\alpha_j - 1/2\,\partial_j(\gamma_{rs}\beta^r\beta^s)] - \beta^k\widehat{\Gamma}^0_{00}$	$\widehat{\Gamma}^k_{ij} = \Gamma^k_{ij} - \beta^k\widehat{\Gamma}^0_{ij}$
$\widehat{\Gamma}^k_{i0} = -\alpha\, K_i{}^k + \nabla_i\beta^k - \beta^k\widehat{\Gamma}^0_{i0}$	$\widehat{\Gamma}^0_{ij} = -1/\alpha\, K_{ij}$

Let us note, however, that 3+1 decompositions became popular from the work of Arnowitt, Deser, and Misner (ADM) about the Hamiltonian formalism [5], and they are often referred to as ADM equations for that reason, although the version appearing in [5] is an equivalent system in which the extrinsic curvature K_{ij} is replaced by the 'conjugate momentum' combination

$$\Pi_{ij} = K_{ij} - trK\,\gamma_{ij} \tag{2.55}$$

as a basic dynamical object. We will refer instead to (2.51), (2.52), (2.53), and (2.54) as the 3+1 field equations, emphasizing in this way the purely geometrical aspects of this approach.

The time-dependent space coordinates transformation (2.43), when applied to the line element (2.22), transforms it to the general form

$$ds^2 = -\alpha^2 dt^2 + \gamma_{ij}\,(dy^i + \beta^i dt)\,(dy^j + \beta^j dt)\,, \tag{2.56}$$

where it is clear that the new time lines $y =$ constant are no longer orthogonal to the constant time slices (see Fig. 2.3). The decomposition (2.56) is actually the most general one, where the four-coordinate degrees of freedom are represented by the lapse α and the shift β^k, whereas the normal coordinates form (2.22) is recovered only in the vanishing shift case.

Using a non-zero shift is certainly a complication. For instance, the inverse matrix of the 4D metric is given by

$$\hat{g}^{00} = -\frac{1}{\alpha^2}\,, \qquad \hat{g}^{0i} = \frac{1}{\alpha^2}\,\beta^i\,, \qquad \hat{g}^{ij} = \gamma^{ij} - \frac{1}{\alpha^2}\,\beta^i\,\beta^j\,, \tag{2.57}$$

and the connection coefficients contain now much more terms (see Table 2.2).

There are physical situations, however, in which a non-zero shift can be very convenient, for instance:

- When rotation is an important overall feature (spinning black holes, binary systems, etc.). If we want to adapt our time lines to rotate with the bodies, then we cannot avoid vorticity and normal coordinates can no longer be used. The shift choice will be then dictated by the overall motion of our system, so that our space coordinates will rotate with the bodies (co-rotating coordinates).

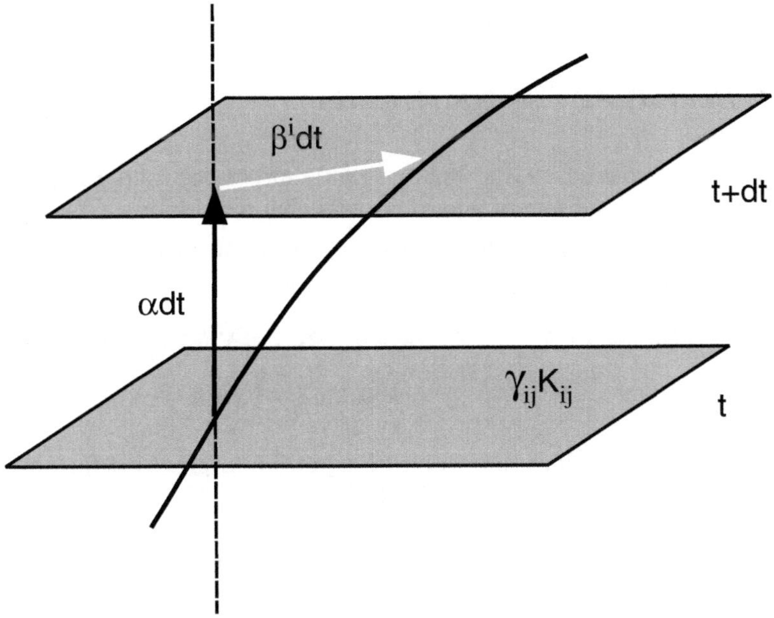

Fig. 2.3 Starting from a given *time slice*, we show the *normal line* (*dashed*) and the *time line* (*continuous*) passing through a given point. The lapse function provides the amount of proper time elapsed when moving to the next *time slice* along the *normal lines*. The shift measures the deviation between these *normal lines* and the actual *time lines* in the process.

- When one needs to use spacelike ('tachyon') time lines. As discussed before, this is allowed provided that the constant time slices remain spacelike. But one cannot have both things in normal coordinates: the squared norm of the vector $\xi^\mu = \delta^\mu_0$, tangent to the time lines, is given by

$$\xi \cdot \xi = -\alpha^2 + \gamma_{rs}\, \beta^r \beta^s\,, \tag{2.58}$$

so that one would need a superluminal shift to do the job, namely

$$|\beta| > \alpha\,. \tag{2.59}$$

The use of a superluminal shift is mandatory if we want to move a black hole across the numerical domain [6]. This is not just a curiosity: it is rather the cornerstone in the 'moving puncture' approach, currently used in binary black hole simulations [7, 8]. It can also be very useful when performing numerical simulations in the vicinity of a black hole, if we want to prevent the horizon from growing too fast, enclosing all of our numerical grid before we have enough time to properly study the exterior region [9–11]. This is also a key ingredient in current binary black hole simulations based on the generalized harmonic formalism [12].

2.3 The evolution system

2.3.1 Evolution and constraints

The 3+1 decomposition (2.51), (2.52), (2.53), and (2.54) splits Einstein's field equations into two subsets of equations of a different kind:

- **Evolution equations.** These govern the time evolution of the basic dynamical fields $\{\gamma_{ij}\,,\ K_{ij}\}$, that is, (2.51) and

$$(\partial_t - \mathcal{L}_\beta)\,K_{ij} = -\nabla_i\,\alpha_j + \alpha\,[R_{ij} - 2K^2_{ij} + trK\;K_{ij}\,] - 8\pi\alpha\,[\,S_{ij} - \frac{1}{2}\,(trS - \tau)\;\gamma_{ij}\,]\,, \qquad (2.60)$$

  where, as stated before, no evolution equation is provided for any of the kinematical (coordinate gauge) fields $\{\alpha\,,\ \beta^i\}$.
- **Energy and momentum constraints.** These are constraints on the extrinsic curvature components K_{ij} and their space derivatives:

$$\mathcal{E} \equiv \frac{1}{2}\,[\,tr\,R + (tr\,K)^2 - tr(K^2)\,] - 8\pi\,\tau = 0 \qquad (2.61)$$

$$\mathcal{M}_i \equiv \nabla_j\,(K_i{}^j - trK\;\delta_i{}^j) - 8\pi\,S_i = 0\,. \qquad (2.62)$$

  The names of energy and momentum correspond to the matter terms appearing in each equation.

Note that we can always re-combine the equations, leading to different partitions of the full system. The constraint subset (2.61) and (2.62), however, can be univocally characterized as the one in which no time derivative of K_{ij} appears. This is not the case of the evolution subset: the one we have got in (2.60) will be called 'Ricci evolution system,' because it corresponds to the space components of the 4D Ricci tensor, as it was obtained in (2.39). One can use the energy constraint (2.61) to cancel out the energy density τ contribution in (2.60), so that the evolution subsystem will consist now in (2.51) plus

$$(\partial_t - \mathcal{L}_\beta)\,K_{ij} = -\nabla_i\,\alpha_j + \alpha\,[\,R_{ij} - 2K^2_{ij} + trK\;K_{ij} - 8\pi\,S_{ij}\,] - \frac{\alpha}{4}\,[\,tr\,R + (tr\,K)^2 - tr(K^2) - 16\pi\,trS\,]\,\gamma_{ij}\,. \qquad (2.63)$$

The subset (2.51) and (2.63) will be called Einstein evolution system, because it can be obtained from the space components of the 4D Einstein tensor, as can be easily seen from the matter terms appearing there. Although the Ricci and the Einstein evolution systems are not equivalent when considered independently, the complete set formed by any of them plus the energy and momentum constraints is the same one, the individual equations being just combined in different ways. We will make use of this recombination freedom in what follows.

2.3.2 Constraints conservation

A first look at the 3+1 version ((2.60), (2.61), and (2.62)) of the field equations shows its strong resemblance to the non-covariant form ((2.1), (2.2), (2.3), and (2.4)) of Maxwell equations, where there is also a subset of evolution equations ((2.3) and (2.4)) for the basic dynamical fields $\{\boldsymbol{E}, \boldsymbol{B}\}$ and a subset of constraints ((2.1) and (2.2)) on their space derivatives. The question arises whether the time derivative of the constraints, allowing for the evolution equations, would lead to new constraints. In the case of Maxwell equations, one can easily verify that this is not the case by taking the time derivatives of (2.1) and (2.2) and using both the evolution equations (2.3) and (2.4) and the charge conservation equation (2.5). This means that the constraints are first integrals of the full evolution system: they are preserved during time evolution. The 4D version of Maxwell equations (2.8) and (2.9) gives us the key to understand this result: both sides are conserved. The left-hand side is conserved by the antisymmetry of the electromagnetic field tensor $F^{\mu\nu}$, whereas the conservation of the right-hand side amounts to that of the charge-current four-vector J^{μ} (2.10).

In the case of the Einstein equations, the straightforward procedure of taking the time derivatives of the constraints (2.61) and (2.62) and then using the evolution equations (2.60) is impractical, even using an algebraic computing program. One can, however, take advantage of the lesson learned in the Maxwell case and look instead to the 4D form of the field equations (1.26), where again we find that both sides are conserved. The Einstein tensor $G^{\mu\nu}$ on the left-hand side is conserved due to the contracted Bianchi identities (1.19), whereas the conservation of the right-hand side amounts to that of the stress–energy tensor $T^{\mu\nu}$ (1.25). This is the idea that we advanced in the previous chapter, when discussing (1.29).

We will address here this point in more detail. Let us start by deriving the 3+1 version of (1.25). The most convenient way is to follow the standard procedure, that is,

- by computing it first in normal coordinates;
- by expressing the results in terms of 3+1 covariant quantities;
- by using then the standard replacement (2.50) to get the general expression, valid in any coordinate system.

We give just the final result for the stress–energy tensor conservation:

$$\frac{1}{\alpha}\,(\partial_t - \mathcal{L}_\beta)\;\tau + \nabla_j\, S^j = \tau\; trK - 2\, S^j\, \partial_j\, ln\alpha + K_{ij}\, S^{ij} \tag{2.64}$$

$$\frac{1}{\alpha}\,(\partial_t - \mathcal{L}_\beta)\; S_i + \nabla_j\, S_i^{\;j} = S_i\, trK - S_i^{\;j}\, \partial_j\, ln\alpha - \tau\, \partial_i\, ln\alpha\,, \tag{2.65}$$

which is the 3+1 version of the general covariant equation (1.25) (remember that all the indices are raised and lowered here with the induced metric γ_{ij}).

In the same way, we can also translate into the 3+1 language the corresponding conservation equation (1.29) for the difference between the left-hand side and the right-hand side of Einstein's field equations. Allowing for (2.61) and (2.62), we get

$$\frac{1}{\alpha}\,(\partial_t - \mathcal{L}_\beta)\;\mathcal{E} + \nabla_j\,\mathcal{M}^j = \mathcal{E}\; trK - 2\,\mathcal{M}^j\,\partial_j\,ln\alpha + K_{ij}\,\mathcal{P}^{ij} \tag{2.66}$$

$$\frac{1}{\alpha}\,(\partial_t - \mathcal{L}_\beta)\;\mathcal{M}_i + \nabla_j\,\mathcal{P}_i^{\;j} = \mathcal{M}_i\, trK - \mathcal{P}_i^{\;j}\,\partial_j\,ln\alpha - \mathcal{E}\,\partial_i\,ln\alpha\,, \tag{2.67}$$

where we have noted

$$\mathcal{P}_{ij} \equiv G_{ij} - 8\pi\,T_{ij}\,. \tag{2.68}$$

Now we can see how any eventual deviation from the energy and momentum constraints $\{\,\mathcal{E}\,,\mathcal{M}_i\}$ would propagate, assuming that the time evolution is given by the Einstein system (2.63):

$$\mathcal{P}_{ij} = 0 \tag{2.69}$$

$$\frac{1}{\alpha}\,(\partial_t - \mathcal{L}_\beta)\;\mathcal{E} + \nabla_j\,\mathcal{M}^j = \mathcal{E}\; trK - 2\,\mathcal{M}^j\,\partial_j\,ln\alpha \tag{2.70}$$

$$\frac{1}{\alpha}\,(\partial_t - \mathcal{L}_\beta)\;\mathcal{M}_i = \mathcal{M}_i\, trK - \mathcal{E}\,\partial_i\,ln\alpha\,. \tag{2.71}$$

The corresponding result for the Ricci evolution equation (2.60) can be obtained in an analogous way, by substituting the corresponding condition

$$\mathcal{P}_{ij} = \mathcal{E}\;\gamma_{ij} \tag{2.72}$$

into the full system (2.66) and (2.67). Independent of the choice, the resulting expression will be a linear homogeneous system on $\{\,\mathcal{E}\,,\mathcal{M}_i\}$, so that our statement that the vanishing of such quantities provides a set of first integrals of the evolution equations holds true, as anticipated from the 4D version (1.29).

2.3.3 Evolution strategies

The structure of the 3+1 field equations (2.60), (2.61), and (2.62) is so similar to that of the Maxwell equations (2.1), (2.2), (2.3), and (2.4) that one can get some inspiration for the equation-solving strategies in electromagnetism in order to do the same in the gravitational case. One can start by solving the constraint equations (2.61) and (2.62) to compute up to four of the six dynamical degrees of freedom (represented here by the components of the extrinsic curvature K_{ij}). When the constraints are first integrals of the evolution system, the evolution equations (2.60) can be used later for computing the two remaining dynamical degrees of freedom.

This 'constrained evolution' approach is specially convenient in astrophysical scenarios where the general relativistic effects can be described as lower order corrections to the Newtonian gravity ones. This is because Newtonian gravity is completely analogous to electrostatics, in the sense that the time evolution of the fields is not provided by the equations. The constraints (2.61) and (2.62) contain then all the Newtonian effects whereas genuine relativistic effects, like the field dynamics leading to gravitational waves, must be found instead in the evolution subset (2.60). One could say that the constraints contain all the dynamical degrees of freedom, apart from the two of them corresponding to gravitational radiation and the ones related with the coordinate gauge freedom, as we will justify later.

From the numerical relativity point of view, the constrained evolution approach, although it can be useful to deal with specific physical situations, is not very convenient for building a general purpose code. There are many reasons for this:

- Constraint equations (2.61) and (2.62) are of elliptic type (exemplified by the Laplace equation). This means that particular solutions are of a non-local nature: they depend strongly on boundary conditions and any local perturbation spreads immediately all over the numerical domain. Spectral methods are specially suited for elliptic equations: they allow to put the outer boundary very far away, even at infinity, where one can set up very reliable boundary conditions, and they usually provide smooth and accurate solutions without consuming too much computational resources.
- Evolution equations (2.60), on the contrary, are more close to the hyperbolic type (exemplified by the wave equation), in the sense that local perturbations propagate over the numerical domain with some finite characteristic speed. This allows the appearance of non-smooth profiles, even weak solutions, that are hard to deal with using spectral methods. Either finite difference or finite volume methods are most commonly used with hyperbolic equations, excepting the cases in which only smooth profiles are expected to appear.
- There is no generic way of algebraically splitting the dynamical degrees of freedom in order to single out the ones corresponding to gravitational radiation. As we will see in the next section, such an algebraic splitting can only be done if one knows in advance the gravitational waves propagation direction. This could be the case in highly symmetrical cases: the gravitational radiation degrees of freedom can be actually neglected in the spherical case, so that the radial direction could be used in problems with approximate spherical symmetry. But there is no such rule for the generic case that could be used in a general purpose numerical code.

The obvious alternative to the constrained evolution approach is to use just the six evolution equations (2.60) to compute everything. The constraints

(2.61) and (2.62) could be enforced just on the initial and boundary data, because they are first integrals of the evolution system and this would be enough to ensure their validity inside the computational domain. The constraints provide just a quality check of the calculations. One can even add some 'damping terms,' devised for minimizing constraint violations [13]. This approach, named 'free evolution' [14], continues to be by far the most commonly used in numerical relativity codes, usually implemented with either finite differences or finite elements discretization. This approach has very deep theoretical and practical implications, which will be discussed thoroughly in the next chapter.

2.4 Gravitational waves degrees of freedom

Gravitational waves constitute one of the most outstanding predictions of Einstein's general relativity. One can derive it by standard methods, in the same way as electromagnetic radiation can be derived from Maxwell's equations. The simplest way is to study small field perturbations around a fixed background, along the lines discussed in Sect. 1.4.2. Relativity textbooks usually assume the harmonic gauge, taking the relaxed system (1.65) as the starting point. This is a little bit misleading, because in that way all metric components seem to generate waves, propagating with light speed. Both the harmonic constraints and the residual gauge freedom must be used in order to isolate the true (just two) gravitational wave degrees of freedom.

We will rather follow here a 3+1 approach, so that both the gauge effects and the constraint degrees of freedom are more easily identified. In this way, we will also gain some insight on the structure of the field equations.

2.4.1 Linearized field equations

Let us come back to the full set (evolution plus constraints) of 3+1 equations (2.51), (2.52), (2.53), and (2.54). To avoid coordinate complications, we will choose here Gauss coordinates, that is, normal coordinates (zero shift) and Geodesic slicing (with $\alpha = 1$), so that

$$ds^2 = -dt^2 + \gamma_{ij}\, dx^i\, dx^j\,. \tag{2.73}$$

As discussed in the previous chapter, any metric can be written down at a given spacetime point P in a locally inertial coordinate system such that the first derivatives of the metric coefficients vanish at P. Then, as we can get as close as we want to P, we can safely split the space metric in (2.73) into two components:

- An Euclidean (flat) background of the form

$$\alpha_{(0)} = 1\,, \qquad \gamma^{(0)}_{ij} = \delta_{ij}\,. \tag{2.74}$$

- A linear perturbation which, when superimposed to the background, allows one to recover the full metric

$$\delta\gamma_{ij} = \gamma_{ij} - \gamma^{(0)}_{ij}\,. \tag{2.75}$$

For further convenience, we will relax here the geodesic slicing condition, allowing also for linear perturbations of the lapse function, namely

$$\delta\alpha = \alpha - \alpha_{(0)}. \tag{2.76}$$

Of course, if the extrinsic curvature can be obtained from the first time derivative of the space metric, one must have for consistency

$$K^{(0)}_{ij} = 0\,, \qquad \delta K_{ij} = K_{ij}\,. \tag{2.77}$$

We can substitute the perturbations (2.75), (2.76), and (2.77) into the 3+1 equations (2.51), (2.52), (2.53), and (2.54) for the vacuum case. We get, up to the linear order

$$\partial_t\,(\delta\gamma_{ij}) = -2\,(\delta K_{ij}) \tag{2.78}$$

$$\partial_t\,(\delta K_{ij}) = -\partial^2_{ij}\,(\delta\alpha) + \delta R_{ij} \tag{2.79}$$

$$0 = \delta^{rs}[\partial_r\,(\delta K_{si}) - \partial_i\,(\delta K_{rs})] \tag{2.80}$$

$$0 = tr\,(\delta R)\,, \tag{2.81}$$

where the trace is computed with the flat background metric (2.74) and the linear order expression for the Ricci tensor is given by

$$\delta R_{ij} = -\frac{1}{2}\,\delta^{rs}[\partial^2_{rs}(\delta\gamma_{ij}) + \partial^2_{ij}(\delta\gamma_{rs}) - \partial^2_{ir}(\delta\gamma_{js}) - \partial^2_{jr}(\delta\gamma_{is})]\,. \tag{2.82}$$

2.4.2 Plane-wave analysis

In order to fully analyze the linear system ((2.78), (2.79), (2.80), and (2.81)), it is convenient to Fourier transform the local perturbation and look at the behavior of a generic plane-wave component, propagating along a given direction n_i, namely

$$\delta\alpha = e^{i\,\omega\cdot x}\,a(\omega, t) \tag{2.83}$$

$$\delta\gamma_{ij} = e^{i\,\omega\cdot x}\,h_{ij}(\omega, t) \tag{2.84}$$

$$\delta K_{ij} = e^{i\,\omega\cdot x}\, k_{ij}(\omega, t)\,, \tag{2.85}$$

where $\omega_k = \omega\, n_k$, $\delta^{\,ij}\, n_i\, n_j = 1$.

Now we can translate the partial differential equations system (2.78), (2.79), (2.80), and (2.81) into the following ordinary differential equations system:

$$\partial_t\; h_{ij} = -2\; k_{ij} \tag{2.86}$$
$$\partial_t\; k_{ij} = \omega^2/2\; [\, h_{ij} - n_i\, h_{nj} - n_j\, h_{ni} + (tr\, h + 2\, a)\, n_i\, n_j\,] \tag{2.87}$$
$$0 = k_{ni} - n_i\, tr\, k \tag{2.88}$$
$$0 = tr\, h - h_{nn}\,, \tag{2.89}$$

where the symbol n replacing an index means contraction with n_i. It is then useful to decompose the Fourier modes into longitudinal (aligned with the propagation direction n_i) and transverse components (tangent to the wave-fronts, which are the surfaces orthogonal to n_i). One gets then three different types of modes, according to their time evolution:

- Three **static** modes, as we get from (2.88):

$$\partial_t\; (h_{ni} - n_i\, tr\, h) = -2\; (k_{ni} - n_i\, tr\, k) = 0. \tag{2.90}$$

- One **gauge** mode, whose evolution is fully determined by lapse perturbations,

$$\partial_t\; tr\, h = -2\; tr\, k\,, \qquad \partial_t\; tr\, k = \omega^2\, a\,, \tag{2.91}$$

  where we have used (2.89).
- Two **wave** modes, oscillating with the Fourier frequency ω,

$$\partial_t\, h_{\perp\perp} = -2\; k_{\perp\perp}\,, \qquad \partial_t\, k_{\perp\perp} = \omega^2/2\; h_{\perp\perp}\,, \tag{2.92}$$

  where the symbol $\perp$ replacing an index means the projection orthogonal to n_i. There are only two independent modes in (2.92), because the trace part vanishes if we allow for (2.89), that is,

$$tr\,(h_{\perp\perp}) = tr\;h - h_{nn} = 0\,. \tag{2.93}$$

  This static mode was already included in (2.90).

Let us focus in the gauge mode (2.91) for a while. In the Gauss coordinate system no lapse perturbations are allowed ($a = 0$). This means that the equations allow then for a linear growth of the trace $tr\, h$ of the metric perturbation, which corresponds to the linear term of the space metric determinant, namely

$$\gamma \equiv det(\gamma_{ij}) \simeq 1 + e^{i\,\omega\cdot x}\,(tr\, h + \cdots), \tag{2.94}$$

so that the first equation in (2.91) corresponds to the evolution of the space volume element $\sqrt{\gamma}$

$$\partial_t \sqrt{\gamma} = -\alpha \, tr \, K, \tag{2.95}$$

and the linear gauge mode corresponds then to an overall expansion (or collapse) of the space metric. This is one of the main reasons why geodesic slicing is not suitable for numerical simulations, because discretization errors coupled with the gauge mode can produce an artificial linear growing which is easily amplified by the non-linear terms, leading to a numerical blowup. This is why we have relaxed this condition to allow for generic gauge perturbations.

A much suitable choice in this context would be instead the 'maximal slicing' condition [15],

$$tr \, K = 0 \, , \tag{2.96}$$

which ensures that the gauge modes are also static ($\partial_t \, tr \, h = 0$). The time evolution of $tr \, K$ is given by (2.41), so that the maximal slicing condition is preserved if and only if the lapse function verifies the consistency condition

$$\frac{1}{\alpha} \, \triangle \alpha = tr(K^2) + 4\pi \, (trS + \tau) \, , \tag{2.97}$$

which reduces to the Laplace equation, up to the linear order, in the vacuum case.

Maximal slicing has been widely used in numerical relativity codes [16–18], leading to smooth and stable lapse profiles at the cost of solving the elliptic (Laplace-like) equation (2.97) at every time step. In the non-vacuum case, (2.97) reduces at the linear order to the Poisson-like equation

$$\triangle(\delta\alpha) = 4\pi \, \rho \, , \tag{2.98}$$

where we have considered the mass density ρ as the first-order contribution to the energy density τ (this amounts to considering kinetic and pressure effects as higher order terms). This is precisely the field equation in Newtonian gravity

$$\triangle \, \phi = 4\pi \, \rho \, , \tag{2.99}$$

which determines the gravitational potential ϕ for a given mass distribution.

It follows that the lapse perturbation $\delta\alpha$ can be identified with the gravitational potential ϕ in the Newtonian limit. This is consistent if we define the Newtonian limit by the following two conditions:

- We consider perturbations of the Minkowski background up to the linear order.
- We ignore any evolution effect (apart from the ones produced by the motion of the sources): this means neglecting gravitational waves, but also enforcing maximal slicing, as we have seen.

This explains why the maximal slicing condition (2.97) is so effective in providing smooth lapse profiles, independently of the riddles produced by time evolution in other field components.

2.4.3 Gravitational waves and gauge effects

Let us summarize now some results from the previous analysis. If we use normal coordinates with maximal slicing, then, up to the linear order, the only dynamical effects on every Fourier component are the ones described by the two transverse traceless degrees of freedom given by

$$\partial_t < h_{\perp\perp} >= -2 < k_{\perp\perp} >, \qquad \partial_t < k_{\perp\perp} >= \omega^2/2 < h_{\perp\perp} >, \tag{2.100}$$

where we have noted by $< \cdots >$ the traceless part, for instance,

$$< k_{\perp\perp} > \equiv k_{\perp\perp} - \frac{1}{2} tr\,(k_{\perp\perp})\,\delta_{\perp\perp}\,. \tag{2.101}$$

Equations (2.100) imply that the dynamical behavior of the selected Fourier component can be described as the one of a plane wave propagating with the speed of light along the selected direction n_i, that is,

$$\delta\, K_{ij} \sim e^{i\,\omega(n\cdot x\,\pm\,t)}\,. \tag{2.102}$$

From the physical point of view, it follows that gravitational waves should be transverse and traceless and should propagate with light speed. One could wonder whether the fact that gravitational radiation consists of two degrees of freedom could be anticipated by the following naive balance: six components in K_{ij} minus four constraints give precisely two 'gravitational radiation' components.

The fallacy in this argument can be easily discovered if one tries to apply it to Maxwell equations (2.1), (2.2), (2.3), and (2.4). There we have six components in the electric and magnetic fields, minus two constraints, so that four components are left. But electromagnetic radiation has only one degree of freedom. This means that there are non-radiative dynamical contributions to the electromagnetic field that contribute to the linear order. The true balance should read six electromagnetic field components minus two constraints give four dynamical degrees of freedom, but only one of them is of a radiative type. In Einstein's equations, instead, non-radiative dynamical effects do not show up at the linear order, where we just find gravitational radiation, aside from eventual gauge effects.

We have found one such gauge effect: the linear mode (2.91) that appears when using geodesic slicing. Another kind of gauge effect would show up when the time coordinate is given by a harmonic function (harmonic slicing), that is,

$$\Box\, x^0 = 0 \;\longleftrightarrow\; \widehat{\Gamma}^{\,0} = 0\,. \tag{2.103}$$

In normal coordinates (zero shift), this amounts to

$$\partial_t\,(\alpha/\sqrt{\gamma}) = 0\,. \tag{2.104}$$

The gauge perturbation $\delta\alpha$ is then dynamically related with the space volume perturbation. In Fourier space we have

$$\partial_t \; (a - tr\, h) = 0, \tag{2.105}$$

and the gauge mode (2.91) can be written as

$$\partial_t \; a = -tr\, k\,, \qquad \partial_t \; tr\, k = \omega^2\, a\,, \tag{2.106}$$

which reproduces the same propagation behavior as that of gravitational waves

$$\delta\alpha \sim e^{i\,\omega(n\cdot x\,\pm\,t)}\,. \tag{2.107}$$

These 'gauge waves' do not describe any physical effect: they are rather an artifact of the gauge choice. We have introduced the harmonic slicing here mainly for two reasons:

- It provides an oscillatory gauge behavior that is between the linear one of geodesic slicing and the static one of maximal slicing. We will use a generalization of this condition in the next chapter in order to obtain hyperbolic evolution systems with a view to numerical simulations.
- The direct relationship (2.103) between the lapse function and the space volume element can be used to avoid collapse singularities in numerical simulations of black hole spacetimes (singularity avoidance, see Fig. 2.2). As we will see later, however, the singularity avoidance behavior of the harmonic slicing gauge is just marginal, so that some stronger singularity-avoidant condition is required for these extreme gravitational collapse scenarios.

References

1. G. Darmois, *Les equations de la Gravitation Einsteinnienne*, Memorial des Sciences Mathematiques **25**, Gauthier-Villars, Paris (1927).
2. A. Lichnerowicz, J. Math. Pures et Appl. **23**, 37 (1944).
3. Y. Choquet-Bruhat, J. Rat. Mec. Anal. **5**, 951 (1956).
4. A. Lichnerowicz, *Studies in the History of General Relativity Series: Einstein Studies*, Vol 3, ed. by J. Eisenstead and A. J. Kox, Birkhäuser, Basel (1992).
5. R. Arnowit, S. Deser and C. W. Misner. *Gravitation: An Introduction to Current Research*, ed. by L. Witten, Wiley, New York (1962). gr-qc/0405109.
6. R. Gomez et al., Phys. Rev. Lett. **80**, 3915 (1998).
7. M. Campanelli, C. O. Lousto, P. Marronetti and Y. Zlochower, Phys. Rev. Lett. **96**,111101 (2006).
8. J. G. Baker et al., Phys. Rev. Lett. **96**, 111102 (2006).
9. M. Alcubierre and B. Brügmann, Phys. Rev. **D63**, 104006 (2001).
10. M. Alcubierre et al., Phys. Rev. **D67**, 084023 (2003).
11. L. Lindblom and M. A. Scheel, Phys. Rev. **D67**, 124005 (2003).
12. F. Pretorius, Phys. Rev. Lett. **95**, 121101 (2005).

13. C. Gundlach, G. Calabrese, I. Hinder and J. M. Martín-García, Class. Quantum Grav. **22**, 3767 (2005).
14. J. Centrella, Phys. Rev. **D21**, 2776 (1980).
15. F. Estabrook et al., Phys. Rev. **D7**, 2814 (1973).
16. L. Smarr and J. W. York, Phys. Rev. **D17**, 1945 (1978).
17. L. Smarr and J. W. York, Phys. Rev. **D17**, 2529 (1978).
18. P. Anninos et al., Phys. Rev. Lett. **71**, 2851 (1993).

Chapter 3
Free Evolution

3.1 The free evolution framework

3.1.1 The ADM system

As we mentioned in the previous chapter, the 'free evolution' approach is by far the most commonly used today in numerical relativity codes. It consists in using just the evolution equations to compute the full set of dynamical quantities (γ_{ij}, K_{ij}). We have seen that the subset of evolution equations is not unique: evolution equations can be modified by adding constraints in many different ways. This implies that we must distinguish among different versions of free evolution, depending on the particular variant of the evolution equation which is selected in each case. The first choice is to take just the space components of the 4D Ricci tensor, namely

$$(\partial_t - \mathcal{L}_\beta)\, \gamma_{ij} = -2\alpha\, K_{ij} \tag{3.1}$$

$$(\partial_t - \mathcal{L}_\beta)\, K_{ij} = -\nabla_i\, \alpha_j + \alpha\, [\, R_{ij} - 2K^2_{ij} + trK\; K_{ij}\,] \\ -8\pi\alpha\, [\, S_{ij} - \frac{1}{2}\, (trS - \tau)\, \gamma_{ij}\,] \tag{3.2}$$

(ADM evolution system).

The constraints are not enforced during evolution, although one can use the differences

$$\mathcal{E} \equiv \frac{1}{2}\, [\, tr\, R + (tr\, K)^2 - tr(K^2)\,] - 8\pi\, \tau = 0 \tag{3.3}$$

$$\mathcal{M}_i \equiv \nabla_j\; (K_i{}^j - trK\; \delta_i{}^j) - 8\pi\, S_i = 0 \tag{3.4}$$

as error indicators in order to monitor the quality of the numerical simulations.

Bona, C. et al.: *Free Evolution*. Lect. Notes Phys. **783**, 49–77 (2009)
DOI 10.1007/978-3-642-01164-1_3

The consistency of the free evolution approach relies in the fact that energy and momentum constraints are first integrals of the evolution system. In the previous chapter, we derived this result (2.66) and (2.67) from the conservation of both the Einstein and the stress–energy tensors. In the specific case of the ADM system (3.1) and (3.2), allowing for (2.72), we get the 'subsidiary system'

$$\frac{1}{\alpha}\,(\partial_t - \mathcal{L}_\beta)\,\mathcal{E} + \nabla_j\,\mathcal{M}^j = 2\,\mathcal{E}\;trK - 2\,\mathcal{M}^j\,\partial_j\,ln\alpha \tag{3.5}$$

$$\frac{1}{\alpha}\,(\partial_t - \mathcal{L}_\beta)\,\mathcal{M}_i + \nabla_i\,\mathcal{E} = \mathcal{M}_i\;trK - 2\,\mathcal{E}\,\partial_i\,ln\alpha\,, \tag{3.6}$$

so that it is clear again that Einstein's solutions will be recovered if and only if the initial data verify

$$\mathcal{E} = 0 \quad \mathcal{M} = 0\,. \tag{3.7}$$

In numerical simulations, however, one must allow for the errors which are inherent to any approximation. This is why we are interested in using a well-posed evolution system. By this we mean that small perturbations of the initial data should not carry us too far away from the original solution, as we have already discussed in Sect. 1.4.1. A well-posed evolution system implies the stability of the solutions at the continuum level, which is a necessary condition for the stability of numerical solutions. In this way, we could ensure that violations of the constraints (3.7) arising during numerical evolution would not grow too fast, so that the numerical solution would be a consistent approximation to the exact Einstein's equations solution, to the required accuracy level.

Note that the same considerations apply *mutatis mutandis* to the harmonic constraints preservation by the relaxed evolution system (1.64), which is well-posed by construction, as its principal part (1.65) consists of an uncoupled set of general relativistic wave equations.

3.1.2 Extended solution space

Let us analyze the situation in more detail. It is clear that the free evolution approach implies using an extended solution space, even at the continuum level: less equations to be fulfilled means more solutions. As a consequence, Einstein's solutions span just a subset of some extended solution space. This subset is characterized by the conditions (3.7), which amount to enforce energy and momentum constraints. This is fully consistent because the subsidiary system (3.5) and (3.6) ensures that the constraints (3.7) are preserved, so that every true Einstein's solution remains so during time evolution. As a consequence, there is a clear-cut separation between the original solutions and the extended ones.

The situation is depicted in Fig. 3.1. The fact that the ADM system (3.1) and (3.2) be well-posed would ensure the stability of the extended set solutions (represented as a rectangle). But true Einstein's solutions, being a subset of the extended ones, would also be stable as a result. This means that they would remain inside their constrained subset (represented as a circle). Note also that the extension is not unique: we could choose (2.63) (Einstein evolution system) instead of (3.2) (Ricci evolution system) for the time evolution of the extrinsic curvature, and we would get as a result a different extension of Einstein's original solution space.

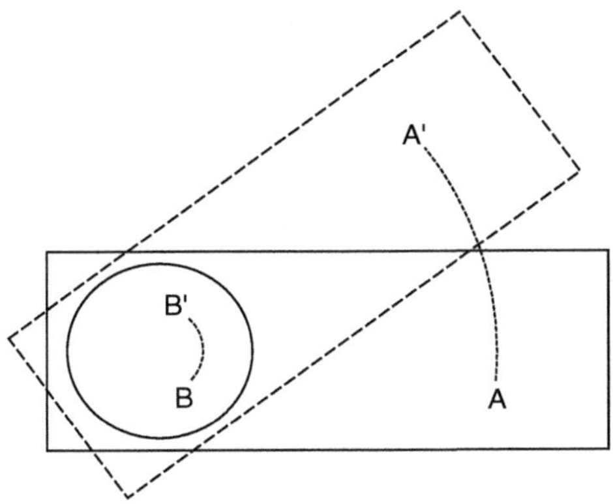

Fig. 3.1 The original Einstein's equations solution space is represented by a *circle*, which is contained into the *rectangle* representing the extended solution space of the ADM free evolution system. The *dashed rectangle* represents a different extension of the original solution space. Stability of the ADM solutions would mean that the computed solutions would not drift far away from the exact ones during time evolution. Then, ADM-extended solutions (point A, for instance) would still be solutions after some time, to the required level of accuracy (they will not drift outside the rectangle). This would imply also that true Einstein's equations solutions (point B, for instance) would remain true solutions, to the required level of accuracy (they will not drift outside the *circle*). The *circular dotted lines* show the effect of a general coordinate transformation, represented here as a mapping of point A into point A′, which is outside the ADM-extended solution space. Point B is mapped instead into B′, which still represents a true solution.

From a different point of view, we must remark that the extended solution space is no longer invariant under general coordinate transformations. Remember that the complete 3 + 1 version of Einstein's equations has the same solution space of the 4D version. This means that a general coordinate transformation maps solutions into solutions, in spite of the fact that the equations themselves are covariant only under the restricted set of transformations (2.43) and (2.44) (3 + 1 covariance).

The ADM system is still 3 + 1 covariant, but remember again that we have obtained it by keeping only the space components of the 4D equations and neglecting the remaining ones. This way we have broken the underlying invariance of (the full set of) Einstein's equations, so that general coordinate transformations will no longer map extended solutions into solutions, as depicted in Fig. 3.1.

3.1.3 Plane-wave analysis

Let us now apply to the ADM system (3.1) and (3.2) the plane-wave analysis, as we did in the previous chapter for the complete $3+1$ system. But we will first generalize it with a view to further applications:

- The uniform background will be allowed now to take (constant) values different from the Minkowski ones, that is,

$$ds^2 = -\alpha_0^2\, dt^2 + \gamma^0_{ij}\, dx^i\, dx^j\,. \tag{3.8}$$

  This means that index raising and lowering will be made using γ^0_{ij}, which can now be different from δ_{ij}. As a consequence, light rays propagating in the background along a given direction n_i travel with coordinate speed (space interval versus coordinate time interval)

$$\frac{dx^i}{dt} = \pm\, \alpha_0\, n^i\,. \tag{3.9}$$

- The Fourier components of the linear perturbations will be rescaled as follows:

$$\delta\alpha = e^{i\,\omega\cdot x}\, a(\omega, t) \tag{3.10}$$

$$\delta\gamma_{ij} = e^{i\,\omega\cdot x}\, h_{ij}(\omega, t) \tag{3.11}$$

$$\delta K_{ij} = (i\omega)\, e^{i\,\omega\cdot x}\, k_{ij}(\omega, t)\,, \tag{3.12}$$

  where we must notice the $i\omega$ factor in the last definition. This is because we will look for wave propagation behavior along the chosen direction n_i, that is,

$$e^{i\,\omega(n\cdot x\, -\, v\, t)}\,, \tag{3.13}$$

  so that the $i\omega$ factor arises from (3.1) defining the extrinsic curvature in terms of the metric.
- The harmonic slicing condition (2.103) will be generalized to

$$\partial_t\, ln\, \alpha = -\alpha\, f\, tr\, K \tag{3.14}$$

  (generalized harmonic slicing [1]), where f is an extra arbitrary factor, so that geodesic slicing is recovered for $f = 0$ and the original harmonic slicing condition is recovered for $f = 1$.

One can now substitute the perturbations (3.10), (3.11), and (3.12) into the linearized ADM system plus the coordinate condition (3.14). This is straightforward if one allows for the expression (2.82) for the 3D Ricci tensor perturbations. In the vacuum case, one gets

$$\partial_t\,(a/\alpha_0) = -i\omega\alpha_0\, f\, tr\, k \tag{3.15}$$

$$\partial_t\, h_{ij} = -2\, i\omega\alpha_0\, k_{ij} \tag{3.16}$$

$$\partial_t\, k_{ij} = -i\omega\alpha_0/2\, [\, h_{ij} - n_i\, h_{nj} - n_j\, h_{ni} + (tr\, h + 2\, a/\alpha_0)\, n_i\, n_j\,] \tag{3.17}$$

and we will write down this linear system in matrix form:

$$\partial_t\, \mathbf{u} = -i\,\omega\, \mathbf{A}\, \mathbf{u}\,, \tag{3.18}$$

where $\mathbf{u}$ is the array of (the Fourier components of the) perturbations:

$$\mathbf{u} = (a\,,\; h_{ij}\,,\; k_{ij})\,. \tag{3.19}$$

The geometric properties of the matrix $\mathbf{A}$ (characteristic matrix) are obviously related with the dynamics of the perturbations:

- The eigenvectors of $\mathbf{A}$ describe modes evolving in time as plane waves (3.13), provided that their corresponding eigenvalues are real. Notice, however, that every physical mode will affect some component of both the metric and its time derivative (the extrinsic curvature), so that it will consist of two eigenvectors.
- Every (real) eigenvalue provides the propagation speed v of the corresponding characteristic mode (characteristic speeds).

The fact that one has a complete set of eigenvectors corresponding to real eigenvalues is important from the physical point of view, because it means that all the modes are wavelike, so that they keep bounded during time evolution. A first-order system verifying this would be said to be 'strongly hyperbolic' (or just 'hyperbolic,' see for instance [2]). The systems we are considering here are of a mixed type: first order in time, but second order in space. This is why we will coin the term 'pseudo-hyperbolic' for mixed-order systems with this property. Also, we will use the term 'weakly pseudo-hyperbolic' in this context when all the eigenvalues are real, but the characteristic matrix $\mathbf{A}$ cannot be fully diagonalized, so that the set of eigenvectors is not complete (the analogous of weak hyperbolicity for fully first-order systems).

In the case of the ADM system, the characteristic matrix can easily be put into a block-diagonal form. One gets from (3.10), (3.11), and (3.12) three uncoupled sectors, containing different types of terms:

- **The transverse sector**, given by

$$\partial_t\, h_{\perp\perp} = -2\, i\omega\alpha_0\; k_{\perp\perp}\,, \quad \partial_t\, k_{\perp\perp} = -i\omega\alpha_0/2\; h_{\perp\perp}\,, \tag{3.20}$$

which consists of six degenerate eigenvectors with light speed $v = \pm\,\alpha_0$ as their characteristic speed.

- **The mixed sector**, given by

$$\partial_t\, h_{n\perp} = -2\, i\omega\alpha_0\; k_{n\perp}\,, \quad \partial_t\, k_{n\perp} = 0\,, \tag{3.21}$$

which consists in a 4D degenerate box with zero as the common eigenvalue, but only two eigenvectors ($k_{n\perp}$), so that the other two ($h_{n\perp}$) can grow linearly with time.

- **The gauge sector**, given by

$$\partial_t\, tr\, h = -2\, i\omega\alpha_0\; tr\, k\,, \tag{3.22}$$

$$\partial_t\, (a/\alpha_0) = -i\omega\alpha_0\, f\, tr\, k\,, \tag{3.23}$$

$$\partial_t\, tr\, k = -i\omega\alpha_0\; a\,, \tag{3.24}$$

which can be fully diagonalized in the generic case, with eigenvalues zero and $v = \pm\sqrt{f}\,\alpha_0$ (gauge speed). The only exception is the $f = 0$ case, which corresponds to geodesic slicing: the gauge sector cannot be fully diagonalized and therefore non-oscillatory growing modes appear, as discussed in the previous chapter.

It follows that the ADM system can be at most weakly pseudo-hyperbolic, when the gauge parameter f is non-negative. But the mixed sector shows that, independent of gauge considerations, the system cannot be pseudo-hyperbolic in the strong sense: the linear growing modes in the mixed sector cannot be avoided. This is in contrast with what we got for the full Einstein system, where either the maximal or the harmonic slicing conditions allowed us to dispose of such annoying modes. This is the price one pays for neglecting the constraints in the free evolution approach. We will see in the next section how high this price can be in numerical simulations and, in the following ones, how one can modify the ADM system (3.1) and (3.2) in order to obtain free evolution pseudo-hyperbolic systems more suitable for building numerical codes.

Could we conclude that the ADM system (3.1) and (3.2) is well posed? Well, it is just a matter of definition. In the mathematical literature, the concept of 'perturbations growing not too fast' is defined in a weak way, so that polynomial growth is admissible [2]. In this sense, the linear modes we have found are not a problem and the ADM system could be well-posed in this weak sense. But we could adopt on physical grounds a stronger requirement so that 'perturbations growing not too fast' is replaced by 'perturbations having an upper bound' and the ADM system cannot be well posed in this strong sense.

3.2 Robust stability test-bed

In this section, we will perform a simple numerical test-bed [3] in order to check out the results of the previous section. The idea is to start from initial data consisting of a flat background (Minkowski metric) plus a random perturbation in every dynamical field. The initial level of the random noise must be small enough to make sure that we are testing just the linear regime, allowing even for the cumulative effects during the time elapsed in the simulation. One should see then a linear growth of the noise level when using the weakly pseudo-hyperbolic ADM system, in contrast with the constant noise level one should get when using any of the pseudo-hyperbolic systems which will be discussed in the following sections.

Besides its use as a cross-check of analytical calculations, the robust stability test-bed can also be useful in two other ways:

- As a direct check of the pseudo-hyperbolic character of a given system, when the analytical calculations are hard to perform. This means using a well-known numerical algorithm, so that any eventual problem will come from the evolution system.
- As a tool for tuning the numerical methods: fixing the time resolution or other adjustable parameters in order to keep stability by avoiding at the same time adding too much numerical dissipation. This means that one is dealing with a well-known strongly hyperbolic system, so that any eventual problem will come from the numerical algorithm.

We will take advantage of both complementary approaches in what follows. But let us first introduce what we consider to be the simplest discretization method, which will be used in our test-bed simulations.

3.2.1 Finite differences

Testing the (pseudo-)hyperbolic character of a given evolution system can be done with very simple numerical algorithms, which one can find in standard books (see for instance [4]). The simplest choice is provided by finite difference methods. The continuous spacetime foliation is approximated by a series of specific time slices, labeled by a time index n. In addition, every space slice, corresponding to a given instant, will be approximated as a 3D grid. The grid nodes will be labeled by a set of three space indices (i, j, k), one for every coordinate axis. The field values will then be represented by the 4D array

$$\mathbf{u}^{(n)}_{i,\,j,\,k} = \mathbf{u}(t_n,\, x_i,\, y_j,\, z_k)\,. \tag{3.25}$$

The quality of this approximation is governed by the size of the time and space intervals, namely

$$\Delta t\,, \qquad \Delta x\,, \qquad \Delta y\,, \qquad \Delta z\,, \tag{3.26}$$

where we are assuming for simplicity that the numerical grid nodes are evenly spaced along every direction.

The space (resp. time) derivatives can be discretized in many ways. For the partial x derivative, for instance, we can take the one-sided differences

$$D^+_x(u) \equiv (u^{(n)}_{i+1,\,j,\,k} - u^{(n)}_{i,\,j,\,k})/\Delta x \qquad D^-_x(u) \equiv (u^{(n)}_{i,\,j,\,k} - u^{(n)}_{i-1,\,j,\,k})/\Delta x\,. \tag{3.27}$$

We can see from a Taylor development that both approximations (3.27) are just first-order accurate, when the leading error is given by the second derivative term (quadratic in Δx).

We can get second-order accuracy for the partial derivatives easily by using centered differences, namely

$$\partial_x\, u \sim D^0_x(u) = \frac{1}{2}\,[D^+_x(u) + D^-_x(u)]\,. \tag{3.28}$$

This 'centering' strategy can also be applied to second partial derivatives in order to get again the required second-order accuracy (third-order leading error term), namely

$$\partial_{xx}\, u \sim D^+_x D^-_x(u) \qquad \partial_{xy}\, u \sim D^0_x D^0_y(u)\,. \tag{3.29}$$

Note that the finite-difference operators (3.27) do commute, like their continuum partial derivative counterparts.

Numerical grids are obviously of a finite size, so that they must begin and end at some point along every direction. For these boundary points, one can no longer use expressions like (3.28) or (3.29), which would require neighbor nodes which may not exist at the outermost grid points. As we are not interested in boundary effects for the moment, we will assume here that our grid has periodic boundaries along every axis (from the geometrical point of view, this implies the topology of a three-torus). Allowing for this, we will assume that the last two nodes at the end of any axis are identical to the first two ones at the beginning, and vice versa. For instance (dropping the time labels for simplicity)

$$u_{N,\,j,\,k} = u_{2,\,j,\,k}\,, \qquad u_{1,\,j,\,k} = u_{N-1,\,j,\,k}\,, \tag{3.30}$$

so that the centered expressions for the first and second partial derivatives (3.28) and (3.29) are applied only for $i = 2, \ldots, N-1$, whereas (3.30) provides the required values at the boundary points $i = 1, N$. We are aware that, in most numerical relativity applications, periodic boundary conditions like

(3.30) would not be consistent with the physics of the problem. But for the moment we are planning to deal just with some test-bed problems, where periodic boundaries could be assumed consistently, in order to focus on the numerical treatment of the generic (interior) points.

In the finite difference approach, the set of grid points needed to discretize space derivatives at a given point P is named as 'the stencil.' The stencil provides the numerical domain of dependence of the selected point P. This means that any perturbation at one of the stencil points will change the computed value at P after a single time step. We can even define the numerical propagation speed as

$$v^i_{\text{num}} = \frac{s\Delta x^i}{\Delta t}\,, \tag{3.31}$$

where s stands here for the stencil size, that is, the maximum number of stencil points besides P along any direction. For instance, we can have $s = 1$ when using the second derivative expressions (3.29), but one could choose instead

$$\partial_{xx}\, u \sim (D^0_x)^2(u)\,, \tag{3.32}$$

which would imply at least $s = 2$ (the actual value will depend on the specific numerical algorithm).

From the physical point of view, when our system describes propagation with some characteristic speeds, the field values at P are causally determined by the values inside the past half-cone with vertex at P, whose slope is given by (the inverse of) the largest characteristic speed of the system. This provides the physical domain of dependence of P. In our case, the largest characteristic speed is either light speed $v = \alpha_0$ or the gauge speed $v = \sqrt{f}\,\alpha_0$ (usually $f > 1$, so that gauge speed is actually the largest one).

As seen in Fig. 3.2, depending on the size of the time step Δt, the physical domain of dependence of P can or cannot be fully contained in the numerical domain of dependence. Consistency requires that the numerical domain of dependence should contain every point that can have a physical influence on P. Otherwise the numerical solution is not allowing for the causal behavior of the dynamical system and this will lead to numerical instabilities, which is the only way our numerical algorithm has to escape from converging to the physical solution.

A necessary condition for numerical stability will be then the Courant condition, stating that the largest characteristic speed v_{max} along every given direction n_i cannot exceed the corresponding numerical speed,

$$v_{\text{max}} \;<\; n_i\, v^i_{\text{num}}\,. \tag{3.33}$$

For instance, let us consider the x direction and let us assume $f \geq 1$; the Courant condition (3.33) provides then an upper limit for the numerical time step, namely

$$\Delta t \;<\; \frac{s\Delta x}{\alpha_0\,\sqrt{f\,\gamma_0^{xx}}}\,. \tag{3.34}$$

In numerical simulations, the time step limit (3.34) must be checked dynamically, at the beginning of every time step. It must be checked at every grid point and along every axis, keeping always the most restrictive upper bound on the time interval. An extra safety factor is also included, so that the upper bound gets even more restrictive. This supplementary factor is usually adjusted by trial and error, either in order to avoid an excess of numerical dissipation or to suppress other unexpected instabilities.

3.2.2 Numerical results

Let us now proceed to the numerical simulation. We will set up a cubic grid of 'only' 50 points along every axis, with periodic boundary conditions. Although we have 50^3 points, the job can be done easily by using a standard personal computer, even a laptop. Space resolution is not an issue here (we will evolve just noise), so that the number of points could be reduced as needed: for instance, we could take advantage of the periodic boundaries condition to set up instead a 'channel' of $50 \times 8 \times 8$ points if we want a quicker response.

We will measure time in 'crossing time' units. This signifies the time it would take for a light ray to make a full trip across the numerical grid. We will choose our time step to be

$$dt = 0.03\, dx\,. \tag{3.35}$$

Notice that the light speed value is one in the background (Minkowski) metric, so that the Courant condition (3.34) would allow us to take a time

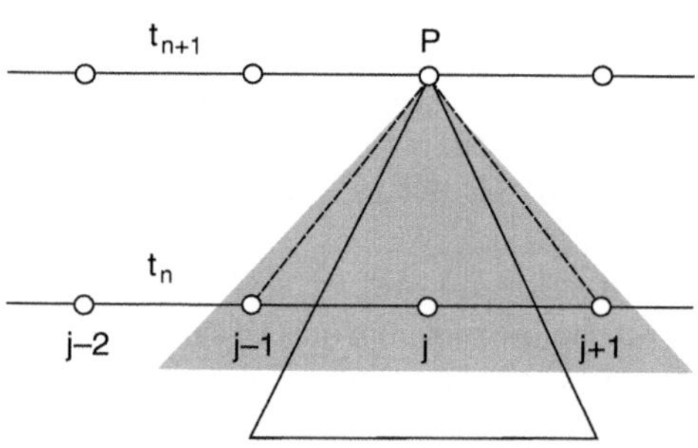

Fig. 3.2 The numerical domain of dependence of a point P in a finite difference grid is limited by *dashed lines*, whose slope is the inverse of the numerical propagation speed (a three-point stencil, $s = 1$, is assumed here). The physical domain of dependence of P is represented as a *cone* (*white* or *gray*), whose slope is the inverse of the largest characteristic speed along the given axis. The Courant stability condition is fulfilled when the physical domain of dependence is fully contained into the numerical one (*white cone*). The converse case (*gray cone*), violating the Courant condition, would lead to numerical instabilities.

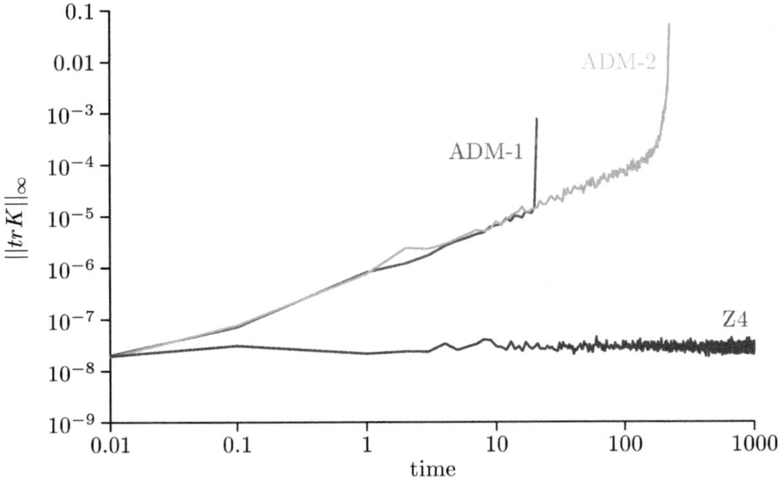

Fig. 3.3 The maximum of (the absolute value of) trK is plotted against the number of crossing times in a logarithmic scale. The initial level of random noise remains constant during the evolution in the case of any of the pseudo-hyperbolic systems that we will describe in what follows (the Z4 one is shown here). In the case of the ADM free evolution system, which is only weakly pseudo-hyperbolic, a linear growth is detected up to the point where the code crashes. The label ADM-1 corresponds to a numerical simulation using a fully first-order version of the ADM system, which will be introduced in the next chapter.

step even 10 times larger. But we prefer to be extremely cautious here about introducing too much numerical dissipation which could artificially lower the noise level, masking the true properties of the evolution system.

We have plotted in Fig. 3.3 our results for the standard harmonic case ($f = 1$). We see the expected linear growth of the ADM system. Notice the catastrophic exponential growth after some 150 crossing times, revealing a non-linear instability. We will limit ourselves to discussing the linear regime as a test for the wave propagation properties of the system. In this sense, the linear growth of the ADM plot in Fig. 3.3 confirms the weakly hyperbolic character of the ADM system.

The Z4 system, which will be introduced later in this chapter, shows instead the constant behavior which one would expect from a strongly pseudo-hyperbolic system. The same qualitative behavior is shown by the BSSN system that will also be introduced later.

Let us briefly discuss the role of numerical dissipation. Every discrete algorithm is just an approximation to the exact equations. Discretization error terms can be classified into two main categories:

- **Dispersion errors**, which affect the propagation speeds. They come from odd-order truncation error terms.
- **Dissipation errors**, which affect the growth of perturbations. They come from even-order truncation error terms.

The wrong sign in the dominant dissipation error terms (perturbations explosion) leads to unstable numerical codes. The right sign there leads to stable codes, at the price of some amount of dissipation. The more accurate the numerical algorithm is, the less numerical dissipation (and dispersion) it contains.

From this point of view, a good deal is provided by third-order-accurate algorithms. This ensures that propagation speeds are correctly represented (fifth-order dispersion error), and the dominant truncation error is represented by a small (fourth-order) dissipation term. We have actually used a third-order Runge–Kutta algorithm for the time evolution in our robust stability test results, as shown in Fig. 3.3. These methods will be explained in more detail in Chap. 5.

3.3 Pseudo-hyperbolic systems

Allowing for the performance of the ADM system in the robust stability test-bed, one can wonder that it was the evolution system of reference until the beginning of the 1990s. Poor resolution, as a consequence of the available computing resources, combined with the use of dissipative numerical methods, masked the weakly stable nature of the formalism (see for instance Fig. 3.4). Also, the linear growing modes (3.21) only show up in truly multidimensional situations. Free evolution with the ADM system was the approach actually used by the main numerical relativity groups in the long way from the pioneering spherically symmetric (1D) or axially symmetric (2D) numerical codes [5–7] toward the fully 3D simulations that are routinely performed today.

3.3.1 Extra dynamical fields

Looking for a better alternative, one can take advantage of the 4D Ricci tensor decomposition (1.61) in order to apply it to its 3D counterpart. The standard expression of the 3D Ricci tensor in terms of the connection coefficients

$$R_{ij} = \partial_k \Gamma^k_{\ ij} - \partial_i \Gamma^k_{\ kj} + \Gamma^k_{\ kr} \Gamma^r_{\ ij} - \Gamma^k_{\ ri} \Gamma^r_{\ kj} \tag{3.36}$$

can then be rewritten, by reordering the second partial derivatives, as

$$\begin{aligned} R_{ij} = \gamma^{rs} \, [-\frac{1}{2} \, \partial^2_{rs} \gamma_{ij} + \partial_{(i} \Gamma_{j)rs} - \Gamma^k_{ij} \, \Gamma_{krs} \\ + \gamma^{kl} \, (\partial_k \gamma_{ir} \, \partial_l \gamma_{js} - \Gamma_{irk} \, \Gamma_{jsl}) \,] \end{aligned} \tag{3.37}$$

(De Donder–Fock decomposition) [8–12].

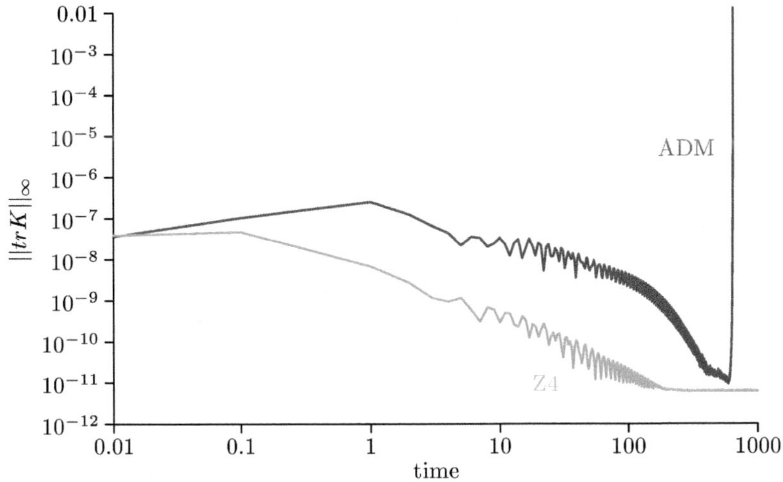

Fig. 3.4 Same as in the previous figure, but replacing the third-order Runge–Kutta time-evolution algorithm by a second-order predictor–corrector one (ICN). Numerical dissipation is severely distorting the plots, by masking the linear growth in the ADM case and dramatically reducing the initial noise level in the pseudo-hyperbolic case. Notice that both dt and dx are the same as in the previous figure, and we are using also the same space discretization algorithm: only the time evolution method has changed.

The second derivative terms in (3.37) are now the sum of the Laplacian of the space metric components plus the symmetrized partial derivatives of the 3D contracted gamma combination

$$\Gamma_i = \gamma^{rs}\Gamma_{irs}\,, \tag{3.38}$$

in the same way as in the 4D version (1.61). The Laplacian term is welcome if one wants to get the expected oscillatory behavior for the linear perturbations. One could be tempted to obtain now the analogous of the relaxed system (1.64), by demanding the vanishing of the quantities (3.38). This would amount to getting 3D harmonic coordinates on every constant time slice. This is actually a very specific choice of the space coordinates, which should then be preserved by time evolution. We will not follow that way, because we prefer to keep our coordinate freedom as longer as possible. Therefore, we must adopt another strategy, devised for working in a generic coordinate system.

The crucial point is to consider instead the quantities Γ_i as the components of a new dynamical field, no longer related with the metric derivatives through (3.38). To be consistent with this interpretation, one must provide an independent evolution equation for this 'extra' field. And this can be done if we allow for the fact that the momentum constraint (3.4) contains space derivatives of the extrinsic curvature, which can be seen itself as a time derivative of the space metric. One can then switch the order of space and time derivatives in (3.4), by rewriting it as a first-order evolution equation for Γ_i [13]. Alternatively, one can express it as a much simpler evolution equation of the combination

$$V_i = \frac{1}{2}\,(\,\Gamma^k_{\ ki} - \Gamma_i\,)\,, \tag{3.39}$$

which is directly related with Γ_i [1]. An equivalent version will be discussed in more detail later in this chapter.

Let us write down here the linearized version of the resulting 'Bona–Massó' system in order to see its structure in a more transparent way (normal coordinates):

$$\partial_t\;(\delta\gamma_{ij}) = -2\alpha_o\;(\delta K_{ij}) \tag{3.40}$$
$$\partial_t\;(\delta K_{ij}) = -\partial^2_{ij}\,(\delta\alpha) + \alpha_o\,(\delta R_{ij}) \tag{3.41}$$
$$\partial_t\;(\delta V_i) = 0\,, \tag{3.42}$$

where the linear perturbation of the Ricci tensor is now given by

$$\delta R_{ij} = -\frac{1}{2}\,\gamma_o^{\ rs}[\,\partial^2_{rs}(\delta\gamma_{ij}) - \partial^2_{ij}(\delta\gamma_{rs})\,] - \partial_i\,(\delta V_j) - \partial_j\,(\delta V_i)\,. \tag{3.43}$$

Note that the perturbations of the extra field V_i are considered here to be independent of the metric and gauge perturbations, so that their evolution equation (3.42) is a genuine part of the free evolution system. This does not

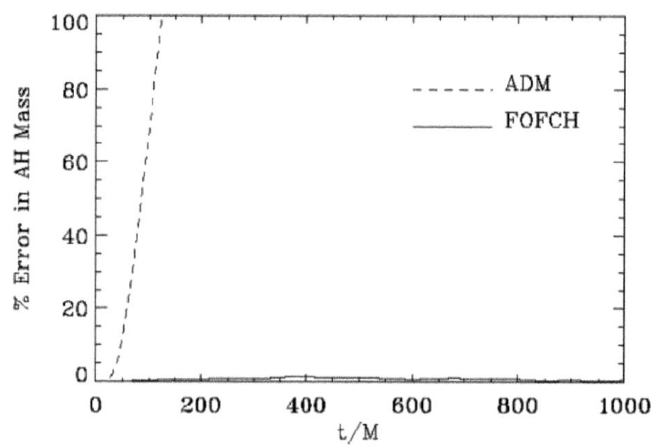

Fig. 3.5 The error in the apparent horizon mass of a spherically symmetric *black hole* is plotted against time. The analytical result is known to be time independent. The term FOFCH, for 'first-order flux-conservative hyperbolic,' refers here to a 1D code based on (the first-order version of) the Bona–Massó formalism [1]. The use of advanced CFD techniques, which require a full set of characteristic eigenvectors, when combined with the use of robust slicing conditions, allows to continue the *black hole* evolution 'forever,' provided that the boundary conditions are set in a suitable way.

mean that the momentum constraint (3.4) has disappeared. It has only been transformed into the algebraic form (3.39), that is,

$$V_i = \frac{1}{2}\gamma^{\,rs}[\,\partial_i\,(\gamma_{rs}) - \partial_r\,(\gamma_{is})\,]\,, \tag{3.44}$$

which can still be used to monitor the quality of the simulation.

The Bona–Massó evolution system was first obtained in the context of fully first-order systems, which we will discuss in the next chapter. In the first (1992) paper [13], it was shown to be hyperbolic in that context for the harmonic slicing case. This result was later extended to the generalized harmonic case [1], which includes the robust slicing conditions currently used at that time to deal with spherically symmetric black holes. It opened the way to the use in the numerical relativity context of the 'shock-capturing' algorithms currently used in computational fluid dynamics (CFD). These advanced techniques, which we will describe in Chap. 5, produced a breakthrough in the state-of-the-art numerical black hole simulations, as shown in Fig 3.5.

3.3.2 The BSSN system

The key idea of the Bona-Massó system, that is transforming the momentum constraint into an evolution equation for some extra fields, was yet in some earlier works by the group of Nakamura [14, 15]. A more elaborated version led to the formalism introduced later by Shibata and Nakamura [16], much widely known after the work of Baumgarte and Shapiro [17] (BSSN system). The metric coefficients γ_{ij} are there expressed in terms of a conformal metric:

$$\tilde{\gamma}_{ij} = e^{-4\,\phi}\;\gamma_{ij} \tag{3.45}$$

with unit determinant, so that

$$e^{4\,\phi} = \gamma^{1/3} = [det(\gamma_{ij})]^{1/3}. \tag{3.46}$$

The second fundamental form K_{ij} is also decomposed into its trace and trace-free components, namely

$$K = \gamma^{ij}\;K_{ij} \tag{3.47}$$

$$\tilde{A}_{ij} = e^{-4\,\phi}\;(K_{ij} - \frac{1}{3}\;K\;\gamma_{ij})\,. \tag{3.48}$$

The conformal decomposition (3.45) allows one to compute easily the connection coefficients of the conformal metric in terms of the original ones, namely

$$\tilde{\Gamma}^k_{\;ij} = \Gamma^k_{\;ij} - 2\,[\,\delta^k_{\;i}\phi_j + \delta^k_{\;j}\phi_i - \phi^k\gamma_{ij}\,]\,, \tag{3.49}$$

so that one can split the 3D Ricci tensor appearing in the ADM evolution equations (3.2) into the corresponding Ricci tensor of the conformal metric plus some space derivatives of the conformal factor:

$$R_{ij} = \tilde{R}_{ij} - 2\,\tilde{\nabla}_i\phi_j + 4\,\phi_i\phi_j - \tilde{\gamma}_{ij}\,\tilde{\gamma}^{rs}(2\,\tilde{\nabla}_r\phi_s + 4\,\phi_r\phi_s)\,. \tag{3.50}$$

The extra field in the BSSN formalism is provided by the contracted conformal connection, that is,

$$\tilde{\Gamma}^i = \tilde{\gamma}^{rs}\,\tilde{\Gamma}^i_{\ rs} \ = -\tilde{\gamma}^{ij}_{\ ,j}\,, \tag{3.51}$$

which can be related with the extra fields (3.39) of the Bona–Massó formalism as follows:

$$V_i = -\frac{1}{2}\,\tilde{\gamma}_{ij}\tilde{\Gamma}^j + 4\,\phi_j\,, \tag{3.52}$$

so that the conformal Ricci tensor can be finally written as

$$\begin{aligned} \tilde{R}_{ij} = \frac{1}{2}\,\tilde{\gamma}^{rs}\,[-\partial^2_{rs}\tilde{\gamma}_{ij} + 4\,\tilde{\Gamma}^k_{\ r(i}\,\tilde{\Gamma}_{j)ks} + 2\,\tilde{\Gamma}^k_{\ ri}\,\tilde{\Gamma}_{ksj}\,] \\ + \tilde{\gamma}_{k(i}\,\partial_{j)}\tilde{\Gamma}^k + \tilde{\Gamma}^k\,\tilde{\Gamma}_{(ij)k}\,. \end{aligned} \tag{3.53}$$

The full list of BSSN-independent dynamical fields is then given (in the zero shift case) by

$$u = \{\alpha,\ \phi,\ \tilde{\gamma}_{ij},\ K,\ \tilde{A}_{ij},\ \tilde{\Gamma}^i\}\,. \tag{3.54}$$

Up to now, the formulas (3.45) to (3.52) just provide a recombination of the basic dynamical fields, but no new equations. The main differences between the BSSN and the Bona–Massó evolution systems are

- The evolution equation for the trace K of the extrinsic curvature in the BSSN case is not the one that would follow from taking the trace of the evolution equation of K_{ij} in a straightforward way. One must use instead the energy constraint to transform the trace equation, before doing all the replacements, into

$$\partial_t\,K = -\triangle\,\alpha + \alpha\,[\,tr(K^2) + 4\pi\,(trS + \tau)\,]\,. \tag{3.55}$$

  This is another instance of the use of the energy constraint to modify the free evolution system: we did the same in the previous chapter, where we compared the 'Ricci' evolution system (2.60) with the 'Einstein' one (2.63). We will see in the following section how to do it in a more systematic way.
- The conformal decomposition, as used in the BSSN formalism, generates new constraints. It follows from (3.45), (3.46), (3.47), and (3.48) that

$$det(\,\tilde{\gamma}\,) = 1\,, \qquad tr\,\tilde{A} = 0\,. \tag{3.56}$$

These algebraic constraints can be very useful for monitoring numerical errors. The current practice is to 'correct' these errors by rescaling $\tilde{\gamma}_{ij}$ and resetting the trace of $\tilde{A}_{ij}$ to zero from time to time during the calculation, even after every single time step. This 'semi-constrained' approach can affect convergence tests, where the rate of convergence that one could expect from the discretization algorithm can differ from the actual results.

3.3.3 Plane-wave analysis

Now we can proceed to perform the plane-wave analysis that we introduced for the ADM system at the beginning of this chapter. We will do it just for the Bona–Massó case in order to avoid the complications associated with the conformal decomposition. Remember that this decomposition, as well as the selection of one or the other of the related quantities (3.52) as the extra field, is only a rearrangement of the dynamical quantities, so that the intrinsic properties of the evolution system remain unchanged. This means that our results will apply also to the BSSN system, with only a minor difference in the gauge sector, due to the use of the modified equation (3.55) for $tr\,K$.

Let us start then from the linearized system (3.40), (3.41), and (3.42). The Fourier components of the dynamical perturbations will be written as follows:

$$\delta\alpha = e^{i\,\omega\cdot x}\,a(\omega,t) \tag{3.57}$$

$$\delta\gamma_{ij} = e^{i\,\omega\cdot x}\,h_{ij}(\omega,t) \tag{3.58}$$

$$\delta K_{ij} = (i\omega)\,e^{i\,\omega\cdot x}\,k_{ij}(\omega,t) \tag{3.59}$$

$$\delta V_k = (i\omega)\,e^{i\,\omega\cdot x}\,v_k(\omega,t)\,, \tag{3.60}$$

where we have scaled the Fourier coefficients of V_k with the same $i\omega$ factor as the K_{ij} ones.

We can now substitute (3.57), (3.58), (3.59), and (3.60) into the linear system (3.40), (3.41), and (3.42) plus (the linear version of) the coordinate condition (3.14) in order to get the time evolution equations for the array of the Fourier-transformed perturbations,

$$\mathbf{u} = (a\,,\; h_{ij}\,,\; k_{ij}\,,\; v_k)\,. \tag{3.61}$$

Note that we must now use the expression (3.43) instead of (2.82) for the 3D Ricci tensor perturbations. In the vacuum case, we get

$$\partial_t\;(a/\alpha_0) = -i\omega\alpha_0\,f\,tr\,k \tag{3.62}$$

$$\partial_t\;h_{ij} = -2\,i\omega\alpha_0\,k_{ij} \tag{3.63}$$

$$\partial_t\, k_{ij} = -i\omega\alpha_0/2\,\left[\,h_{ij} + 2(n_i\, v_j + n_j\, v_i) + (2a/\alpha_0 - tr\,h)\, n_i\, n_j\,\right] \quad (3.64)$$

$$\partial_t\, v_k = 0\,. \quad (3.65)$$

In order to write down the characteristic matrix, it is convenient to rearrange the dynamical fields array (3.61) in a way that clearly separates the transverse, longitudinal, and gauge sectors, as we did in the ADM case:

$$\mathbf{u} = (h_{\perp\perp}\,,\; k_{\perp\perp}\,,\; h_{n\perp}\,,\; k_{n\perp}\,,\; v_\perp\,,\; trh\,,\; trk\,,\; a/\alpha_0\,,\; v_n)\,. \quad (3.66)$$

The characteristic matrix is given in this basis in a simple block-diagonal form

$$\mathbf{A} = \alpha_0 \begin{pmatrix} 0 & 2 & & & & & & & \\ 1/2 & 0 & & & & & & & \\ & & 0 & 2 & 0 & & & & \\ & & 1/2 & 0 & 1 & & & & \\ & & 0 & 0 & 0 & & & & \\ & & & & & 0 & 2 & 0 & \\ & & & & & 0 & 0 & 1 & 2 \\ & & & & & 0 & f & 0 & 0 \\ & & & & & & & & 0 \end{pmatrix} \quad (3.67)$$

(the values not shown are zero), so it follows that

- The **eigenvalues** (propagation speeds) are either zero (static modes, like the (3.65) ones), or the background metric light speed $\pm\alpha_0$, or the 'gauge speed' $\pm\sqrt{f}\,\alpha_0$. The requirement of real propagation speeds amounts then to the condition $f \geq 0$ on the gauge parameter f.
- A complete set of **eigenvectors** can be obtained in the generic case, with the only exception of geodesic slicing ($f = 0$), where gauge speed vanishes leading to an extra degeneracy which prevents putting $\mathbf{A}$ in full diagonal form.

We can conclude that the Bona–Massó system is pseudo-hyperbolic for all the generalized harmonic slicing cases for which

$$f > 0\,. \quad (3.68)$$

The same is true for the BSSN system, where only minor changes appear in the corresponding characteristic matrix when compared with (3.67). The main one arises from the choice of the evolution equation (3.55) for $tr\,K$, and it amounts to the vanishing of the $2\alpha_0$ coefficient in the last column (gauge sector). Therefore, it does not affect either the values of the propagation speeds or the completeness of the set of eigenvectors.

3.4 Covariant formulations

The second-order pseudo-hyperbolic formalisms we have discussed here represent an improvement over the free evolution ADM system, both at the theoretical and at the numerical applications level, where the pseudo-hyperbolicity property ensures the absence of growing modes in linear perturbations. This fact has proven to be very useful in 3D numerical black hole simulations, using either the Bona–Massó [18] or the BSSN formalisms [19, 20]. In the BSSN case, the combined use of the 'gamma-driver' shift prescription [21] (devised to freeze the $\tilde{\Gamma}^i$ evolution) and the 'moving punctures' approach [22, 23] has allowed to perform long-term simulations of binary black holes systems.

These formalisms, however, share two drawbacks:

- Energy and momentum constraints are treated in a different way. Energy constraint is treated like in the ADM case, by relaxing it during time evolution. Momentum constraint, instead, is considered to provide the evolution equation for some extra dynamical field (V_i or $\tilde{\Gamma}^i$, respectively).
- These extra dynamical quantities have no tensor behavior: neither V_i nor $\tilde{\Gamma}^i$ transforms as three vectors under general space coordinates transformations.

The second point was later corrected, leading to the three-covariant Z3 formalism [24], which will be discussed below. But the first point clearly suggests that all these are just intermediate steps toward more advanced formalisms, in which the energy constraint is not left aside and the supplementary dynamical quantities have a well-defined four-tensor character. All these requirements are fulfilled by the 'Z4 formalism' [25, 26], which we will present in what follows.

3.4.1 The Z4 formalism

The field equations in the Z4 formalism can be written in a general covariant form at the 4D level, namely

$$R_{\mu\nu} + \nabla_\mu Z_\nu + \nabla_\nu Z_\mu = 8\,\pi\,(T_{\mu\nu} - \frac{1}{2}\,T\,g_{\mu\nu})\;, \tag{3.69}$$

where the 'zero' four-vector Z_μ plays the role of the supplementary quantity, so that the full set of dynamical fields consists of the pair

$$\{g_{\mu\nu}, Z_\mu\}\,. \tag{3.70}$$

Here again, the solution space of the original Einstein's equations is extended by introducing the extra dynamical quantity Z_μ. The solutions of the original field equations can of course be recovered by imposing the vanishing of the additional terms, namely

$$\nabla_\mu Z_\nu + \nabla_\nu Z_\mu = 0, \tag{3.71}$$

which is the Killing equation. For a generic spacetime, the only solution of this equation is the trivial one:

$$Z_\mu = 0\,. \tag{3.72}$$

The four components of this algebraic condition will play the role of the energy–momentum constraints, as we will see later. One can even use the evolving values of Z_μ during a numerical simulation as a good general covariant indicator of the quality of the approximation. Note that, in contrast with the precedent formalisms, the extended solution space is also general covariant, so that a generic 4D coordinate transformation maps solutions into solutions for both the original Einstein's equations and the Z4 extended ones (3.69) (see Fig. 3.6).

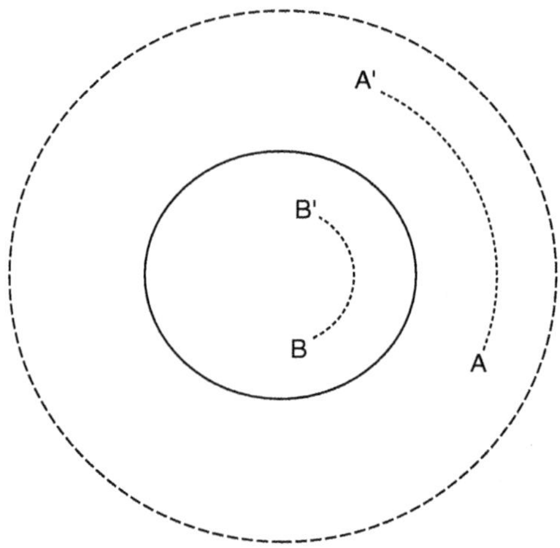

Fig. 3.6 The original Einstein's equations solution space is represented by the *inner circle*, which is contained into the *outer circle* representing the extended solution space of the Z4 system. The plot shows the effect of a general coordinate transformation, represented here as a mapping with *circular dotted lines*. Points A, representing a Z4-extended solution, and B, representing a true Einstein's solution, are mapped into A′ and B′, which still represent either a Z4-extended solution or a true Einstein solution, respectively. This is in contrast with the behavior shown in Fig. 3.1 for the ADM system, reflecting the lack of invariance of the ADM-extended solution space.

The time evolution of the algebraic constraint (3.72) can be obtained by taking the four-divergence of the field equations (3.69) and allowing for the conservation of both the Einstein tensor $G_{\mu\nu}$ and the stress–energy tensor $T_{\mu\nu}$, that is,

$$\nabla_\nu [\, \nabla^\mu Z^\nu + \nabla^\nu Z^\mu - (\nabla_\rho Z^\rho)\, g^{\mu\nu} \,] = 0\,. \tag{3.73}$$

After a straightforward calculation, one gets the equivalent form

$$\Box\, Z_\mu + R_{\mu\nu} Z^\nu = 0\,, \tag{3.74}$$

which plays here the role of the subsidiary system (3.5) and (3.6). It is of second order in Z_μ, so that the vanishing of both Z_μ and its first time derivatives is needed at the initial slice if we want the algebraic constraint (3.72) to be preserved during time evolution. As we will see below, the vanishing of the first derivatives of Z_μ amounts to imposing the original energy and momentum constraints. This means that the initial data set leading to true Einstein's solutions must consist in an initial metric $g_{\mu\nu}$ verifying the usual energy and momentum constraints plus a zero initial value for the four-vector Z_μ.

3.4.2 The generalized harmonic formalism

It is interesting to note at this point that the generalized harmonic system (1.82) and the Z4 system (3.69) can be related by defining the 'gauge sources' H^μ as

$$H^\mu = -\Gamma^\mu - 2Z^\mu\,, \tag{3.75}$$

instead of (1.83).

From the generalized formalism approach, the specification of the gauge sources H^μ is fully equivalent to the specification of the quantity Z^μ in (3.75).

In particular, the relaxed system (1.64) is obtained when imposing

$$Z^\mu = -\frac{1}{2}\,\Gamma^\mu \qquad (H^\mu = 0)\,. \tag{3.76}$$

In this context, it is clear that the harmonic constraints (1.63) amount to the algebraic constraints (3.72), that is,

$$\Gamma^\mu = 0 \qquad \Leftrightarrow \qquad Z^\mu = 0\,. \tag{3.77}$$

This equivalence provides a nice alternative proof of the fact that the harmonic constraints are first integrals of the relaxed system. We give here just an outline:

- The relaxed system (1.64) is precisely the Z4 system for the choice (3.76).
- The vanishing of Z^μ implies the fulfillment of the harmonic constraints (1.63).
- Any four-vector Z^μ satisfying the Z4 system must satisfy the subsidiary system (3.74).
- The vanishing of Z^μ is a first integral of the subsidiary system (3.74).

The preceding arguments could suggest that the Z4 system is just another version of the generalized harmonic formalism. This would be true only if the Z^μ vector is treated as a gauge source that should be specified by some additional prescription. But in the Z4 approach Z^μ is rather an independent dynamical field, defined just by its evolution equations, which are given by the covariant evolution system (3.69).

Let us note, however, that, in some recent generalized harmonic formulations, the gauge sources are also considered as independent dynamical fields (see for instance [27]). In these cases, the formal equivalence with the Z4 formalism is enhanced. Equation (3.75) provides the relationship between the supplementary dynamical fields of the two formulations.

3.4.3 Constraint-violation control

As we will see below, the algebraic constraints (3.72) amount to the standard energy–momentum constraints (2.61) and (2.62). The fact that these constraints get an algebraic form means that the supplementary dynamical fields Z_μ can be used as a quality control, providing a quantitative measure of the energy–momentum constraints violation. The subsidiary system (3.73) is actually providing the time evolution of these constraints violations.

Let us note at this point that the Z4 system is of mixed order: second order in the metric but first order in Z_μ. The subsidiary system (3.73) is instead of second order in Z_μ. We can use then the Z4 field equations to replace the Ricci tensor in (3.74). The principal part of the resulting system

$$\Box\, Z_\mu = [\, \nabla_\mu Z_\nu + \nabla_\nu Z_\mu - 8\,\pi\,(T_{\mu\nu} - \frac{1}{2}\,T\,g_{\mu\nu})\,]\,Z^\nu \tag{3.78}$$

is just a relativistic wave equation, and it follows that energy–momentum constraints violations do actually propagate with light speed.

At first sight, this can seem satisfactory, meaning that small constraint violations arising from poor initial data, boundary conditions, or just numerical truncation errors will propagate out of the computational domain, provided that suitable outflow conditions are set up at the outer boundary. Assuming a consistent approach and good numerical resolution, constraint violations would be just a transient problem, not affecting long-term results. But this is actually not the case in black hole simulations: constraints violations cannot escape from the interior region because this would require a superluminal propagation speed. This can cause these deviations to cumulate and grow inside the horizon, unless we excise the interior region and set a constraint-preserving outflow condition at the resulting inner boundary.

An interesting alternative is to introduce some constraint-damping terms in the field equations, so that extended solutions are driven toward true Einstein's solutions. This idea ('lambda system,' see [28]) can be easily implemented in the Z4 framework, because of the algebraic nature of the constraints (3.72). Following [29], we will modify the original Z4 system (3.69) in the following way:

$$G_{\mu\nu} + \nabla_\mu Z_\nu + \nabla_\nu Z_\mu - \nabla_\rho Z^\rho \, g_{\mu\nu} + \kappa_1 (n_\mu Z_\nu + n_\nu Z_\mu + \kappa_2 \, n_\rho Z^\rho \, g_{\mu\nu}) = 8\pi \, T_{\mu\nu} \,, \tag{3.79}$$

where κ_1 and κ_2 are free parameters and n_μ is the unit normal to the time slices.

The explicit use of the quantity n_μ in the field equations (3.79) provides a privileged time slicing, in which its space components vanish: $n_i = 0$ (the time component value would depend on the slicing parametrization). This can be surprising, as far as we are modifying a four-covariant system. But driving to zero the Z_μ values is not a time-symmetric operation, and breaking the time symmetry of the equations requires using such privileged time coordinate. We will see in the following section that the extra terms in (3.79) provide damping terms on the Z_μ evolution equations, just as required.

As a final remark, let us note that the close relationship between the Z4 and the generalized harmonic system, as described in the preceding subsection, allows one to adapt easily the extra damping terms from (3.79) to the generalized harmonic context (see [29] for details). This mechanism was used successfully by Pretorius in his pioneering binary black hole simulations [27].

3.5 The Z4 evolution system

3.5.1 3 + 1 *Decomposition*

The extended field equations (3.79) can be written in the equivalent 3 + 1 form (Z4 evolution system) [25]

$$(\partial_t - \mathcal{L}_\beta) \, \gamma_{ij} = -2 \, \alpha \, K_{ij} \tag{3.80}$$

$$\begin{aligned}(\partial_t - \mathcal{L}_\beta) \, K_{ij} = & -\nabla_i \alpha_j + \alpha \, [R_{ij} + \nabla_i Z_j + \nabla_j Z_i \\ & -2 \, K^2_{ij} + (trK - 2\Theta) \, K_{ij} - \kappa_1 (1 + \kappa_2) \, \Theta \, \gamma_{ij}] \\ & -8\pi\alpha \, [\, S_{ij} - \frac{1}{2} \, (trS - \tau) \, \gamma_{ij} \,]\end{aligned} \tag{3.81}$$

$$\begin{aligned}(\partial_t - \mathcal{L}_\beta) \, \Theta = & \frac{\alpha}{2} \, [\, tr\, R + 2 \, \nabla_k Z^k + (trK - 2\Theta) \, trK - tr(K^2) \\ & -2 \, Z^k \alpha_k / \alpha - \kappa_1 (2 + \kappa_2) \, \Theta - 16\pi \, \tau \,]\end{aligned} \tag{3.82}$$

$$\begin{aligned}(\partial_t - \mathcal{L}_\beta) \, Z_i = & \, \alpha \, [\, \nabla_j \, ({K_i}^j - {\delta_i}^j trK) + \partial_i \Theta \\ & -2 \, {K_i}^j \, Z_j - \Theta \, \alpha_i / \alpha - \kappa_1 Z_i - 8\pi \, S_i \,],\end{aligned} \tag{3.83}$$

where Θ stands for the projection of the four-vector Z_μ along the unit normal n_μ to the constant time hypersurfaces, namely

$$\Theta \equiv n_\mu Z^\mu = \alpha\, Z^0 \,. \tag{3.84}$$

In the 3+1 form (3.80), (3.81), (3.82), and (3.83), it is evident that the Z4 evolution system consists just in evolution equations. The constraints (3.72) that can be translated into

$$\Theta = 0, \qquad Z_i = 0 \tag{3.85}$$

are algebraic, so that the full set of field equations (3.79) is actually used during evolution. Equations (3.82) and (3.83) govern the evolution of the Z_μ pieces, and we can see there the additional damping terms, devised for driving the generalized solutions toward true Einstein's solutions, provided that they get the right signs, namely [29]

$$\kappa_1 > 0, \qquad \kappa_2 > -1\,. \tag{3.86}$$

This is in contrast to the ADM formalism [30], which can be recovered from (3.80), (3.81), (3.82), and (3.83) by imposing (3.85). The first two equations (3.80) and (3.81) would transform into the ADM free evolution system, whereas the last two equations (3.82) and (3.83) would transform into the standard energy and momentum constraints, that is,

$$tr\, R + tr^2 K - tr(K^2) = 16\pi\, \tau \tag{3.87}$$

$$\nabla_j \, (K_i{}^j - \delta_i{}^j tr K) = 8\pi\, S_i \,. \tag{3.88}$$

These constraints are not enforced in the free evolution approach, and no mechanism is available for driving generalized solutions toward true Einstein's solutions in the ADM case. The same result holds for both the Bona–Massó and the BSSN formalisms, where the energy constraint (3.87) was still not enforced, although the momentum constraint was treated in a way which anticipated (3.83), as we will see later.

The appearance of the new dynamical quantity Θ, which behaves as a scalar under general space coordinate transformations, allows one to generalize the time slicing condition (3.14) by adding a linear coupling with Θ, that is,

$$\partial_t \, ln\, \alpha = -f \alpha \, (tr\, K - m\, \Theta)\,. \tag{3.89}$$

The new gauge parameter m arising here is by no means superfluous, even if it does not change anything for true Einstein's solutions, where Θ vanishes. The non-trivial parameter value $m = 2$ will be required to get a pseudo-hyperbolic system in the standard harmonic case ($f = 1$), as we will see below.

3.5.2 Plane-wave analysis

Let us consider the linearized version of the Z4 system (3.80), (3.81), (3.82), and (3.83) in order to study the propagation of a plane wave in a stationary and homogeneous background (zero shift):

$$\delta\,\gamma_{ij} = e^{i\,\omega\cdot x}\;h_{ij}(\omega,t) \tag{3.90}$$
$$\delta\,\alpha = e^{i\,\omega\cdot x}\;a(\omega,t) \tag{3.91}$$
$$\delta\,K_{ij} = (i\,\omega)\;e^{i\,\omega\cdot x}\;k_{ij}(\omega,t) \tag{3.92}$$
$$\delta\,\Theta = (i\,\omega)\;e^{i\,\omega\cdot x}\;\theta(\omega,t) \tag{3.93}$$
$$\delta\,Z_k = (i\,\omega)\;e^{i\,\omega\cdot x}\;z_k(\omega,t)\,. \tag{3.94}$$

We shall switch off the damping parameters here (by setting $\kappa_1 = 0$) in order to focus the analysis on the wave propagation properties.

The time evolution of the Fourier coefficients is given by

$$\partial_t h_{ij} = -2\,(i\,\omega)\;\alpha_0\;k_{ij} \tag{3.95}$$
$$\partial_t(a/\alpha_0) = -(i\,\omega)\;\alpha_0\;f\;[\,tr\,k - m\,\theta\,] \tag{3.96}$$
$$\partial_t\theta = -\frac{1}{2}\;(i\,\omega)\;\alpha_0\;[\,tr\,h - h_{nn} - 2\,z_n\,] \tag{3.97}$$
$$\partial_t z_i = -(i\,\omega)\;\alpha_0\;[\,n_i\;(tr\,k - \theta) - k_{ni}\,] \tag{3.98}$$
$$\partial_t k_{ij} = -\frac{1}{2}\;(i\,\omega)\;\alpha_0\;\lambda_{ij}, \tag{3.99}$$

where we have noted

$$\lambda_{ij} \equiv h_{ij} + n_i\,n_j\;(tr\,h + 2\,a/\alpha_0) - n_i\;(h_{nj} + 2\,z_j) - n_j\;(h_{ni} + 2\,z_i), \tag{3.100}$$

and where the symbol n replacing an index means the contraction with the unit vector n_i.

In order to write down the characteristic matrix, it is convenient to rearrange the dynamical fields array in a way that clearly separates the transverse, longitudinal, and gauge sectors, as we did before:

$$\mathbf{u} = (h_{\perp\perp}\,,\;k_{\perp\perp}\,,\;h_{n\perp}\,,\;k_{n\perp}\,,\;z_\perp\,,\;trh\,,\;trk\,,\;a/\alpha_0\,,\;\theta\,,\;v_n)\,, \tag{3.101}$$

where the symbol $\perp$ replacing an index means the projection orthogonal to n_i and we have used the shorthand

$$v_n = \frac{1}{2}\;(tr\,h - h_{nn}) - z_n\,. \tag{3.102}$$

The characteristic matrix of the Z4 evolution system is then given in this basis in the simple block-diagonal form:

$$
\mathbf{A} = \alpha_0 \begin{pmatrix}
0 & 2 & & & & & & & & \\
1/2 & 0 & & & & & & & & \\
 & & 0 & 2 & 0 & & & & & \\
 & & 0 & 0 & -1 & & & & & \\
 & & 0 & -1 & 0 & & & & & \\
 & & & & & 0 & 2 & 0 & & \\
 & & & & & 0 & 0 & 1 & 0 & -2 \\
 & & & & & 0 & f & 0 & -mf & 0 \\
 & & & & & & & & 0 & 1 \\
 & & & & & & & & 1 & 0
\end{pmatrix} . \tag{3.103}
$$

A simple inspection of this matrix, by comparing it with (3.67), shows that

- A new sector appears (**energy sector**) at the lower right corner, involving the pair of Fourier coefficients:

$$
(\theta \, , v_n) \, . \tag{3.104}
$$

  The corresponding eigenvalues are given by the background metric light speed $\pm\alpha_0$. There is a non-trivial coupling between the energy and the gauge sectors.
- The **eigenvalues** (propagation speeds) are either zero, or the background metric light speed, or the 'gauge speed' $\pm\sqrt{f}\,\alpha_0$. The requirement of real propagation speeds amounts again to the condition $f \geq 0$ on the gauge parameter f.
- A complete set of **eigenvectors** can be obtained in the generic case, with the only exceptions of geodesic slicing ($f = 0$), where gauge speed vanishes, and the standard harmonic case ($f = 1$), where gauge speed coincides with light speed. In both cases the extra degeneracy is a problem, although in the second one we can still put $\mathbf{A}$ in full diagonal form when $m = 2$.

We can conclude that the Z4 evolution system (3.69) is strongly pseudo-hyperbolic for all the generalized harmonic slicing cases for which

$$
f > 0 \, , \qquad m = 2 \ \text{ if } \ f = 1 \, , \tag{3.105}
$$

so that we can confirm the relevance of the new gauge parameter m in the generalized harmonic condition (3.89).

The role of the additional linear terms appearing in (3.79) can be studied by including them in the linear perturbation analysis. The resulting propagation speeds will get an imaginary part, describing damping instead of just propagation. We refer the reader to [29] for details. Let us just note here that the transverse traceless sector, which is common to (3.67) and (3.103), is unaffected by the damping terms in (3.79). This means that the damping terms are not affecting gravitational waves propagation. The energy–momentum constraints degrees of freedom are affected instead in a way that depends on the specific choice of the damping parameters κ_1, κ_2.

3.5.3 Symmetry breaking

We will address now the question of the relationship of the Z4 system with the precedent ones. To this end, let us consider the following recombination of the dynamical fields:

$$\tilde{K}_{ij} \equiv K_{ij} - \frac{n}{2}\,\Theta\,\gamma_{ij}, \tag{3.106}$$

so that the Z4 system (3.80), (3.81), (3.82), and (3.83) can be written in a one-parameter family of equivalent forms just by replacing everywhere

$$K_{ij} \rightarrow \tilde{K}_{ij} + \frac{n}{2}\,\Theta\,\gamma_{ij}. \tag{3.107}$$

These kinds of transformations leave invariant the solution space of the system (it is actually the same system expressed in a different basis of dynamical fields).

But suppose now that we want to enforce the first algebraic constraint in (3.85), that is,

$$\Theta = 0\,. \tag{3.108}$$

This amounts to suppress the Θ field as a dynamical quantity. If this suppression is made after the replacement (3.107), one gets a one-parameter family of non-equivalent extended systems with only the three components of the vector Z_i as supplementary quantities (Z3 evolution systems [24]), namely

$$(\partial_t - \mathcal{L}_\beta)\,\gamma_{ij} = -2\,\alpha\,K_{ij} \tag{3.109}$$

$$\begin{aligned}(\partial_t - \mathcal{L}_\beta)\,K_{ij} = & -\nabla_i\,\alpha_j + \alpha\,[\,R_{ij} + \nabla_i Z_j + \nabla_j Z_i \\ & -2\,K^2_{ij} + trK\,K_{ij}\,] - 8\pi\alpha\,[\,S_{ij} - \frac{1}{2}\,(trS - \tau)\,\gamma_{ij}\,] \\ & - \frac{n}{4}\,\alpha\,[\,tr\,R + 2\,\nabla_k Z^k + tr^2 K - tr(K^2) - 2\,Z^k \alpha_k/\alpha - 16\pi\,\tau\,]\,\gamma_{ij}\end{aligned} \tag{3.110}$$

$$(\partial_t - \mathcal{L}_\beta)\,Z_i = \alpha\,[\,\nabla_j\,(K_i{}^j - \delta_i{}^j\;trK) - 2K_i{}^j Z_j - \kappa_1\,Z_i - 8\pi\,S_i\,], \tag{3.111}$$

where we have suppressed the tilde over K_{ij}, allowing for the vanishing of Θ.

Concerning the gauge prescription (3.89), we have from (3.106)

$$trK = tr\tilde{K} - \frac{3n}{2}\,\Theta\,. \tag{3.112}$$

This means that the gauge dynamics derived in the Z3 framework from

$$\partial_t\,ln\,\alpha = -f\alpha\,tr\,\tilde{K} \tag{3.113}$$

would actually correspond to the one derived in the Z4 context from (3.89), provided that we take

$$m = \frac{3n}{2} \,. \tag{3.114}$$

The recombination symmetry (3.107) of the original Z4 system (3.80), (3.81), (3.82), and (3.83) is broken in the transition to the Z3 one (3.109), (3.110), and (3.111), which is produced by the vanishing of the Θ parameter. This 'symmetry breaking' means that any two different choices of the n parameter in the family of Z3 evolution systems (3.109), (3.110), and (3.111) are not mutually equivalent: their solution spaces are different, even at the linear order.

The Bona–Massó system [13] can be easily recovered from the $n = 0$ case in (3.109), (3.110), and (3.111). The extra quantities V_i can be obtained from the vector Z_i as follows:

$$V_i = \frac{1}{2} \gamma^{\,rs} [\, \partial_i \,(\gamma_{rs}) - \partial_r \,(\gamma_{is}) \,] - Z_i \,, \tag{3.115}$$

so that the evolution equation for V_i can be computed in a straightforward way. Notice that, contrary to what happened with V_i, the 'zero' vector Z_i behaves as a three vector under general space coordinate transformations. This difference is clearly reflected in the relationship (3.115) between these two quantities.

The BSSN system [16, 17] can also be recovered (not so easily) from the $n = 4/3$ ($m = 2$) case in (3.109), (3.110), and (3.111). To do this, the Z3 system must be decomposed into trace and trace-free parts

$$e^{4\,\phi} = \gamma^{1/3} \,, \quad \tilde{\gamma}_{ij} = e^{-4\,\phi}\, \gamma_{ij} \tag{3.116}$$

$$K = \gamma^{ij}\, K_{ij} \,, \quad \tilde{A}_{ij} = e^{-4\,\phi}\, (K_{ij} - \frac{1}{3}\, K\, \gamma_{ij}) \tag{3.117}$$

$$\tilde{\Gamma}_i = -\tilde{\gamma}_{ik}\, \tilde{\gamma}^{kj}_{\ \ ,j} + 2\, Z_i \tag{3.118}$$

in order to follow the correspondence with BSSN more closely.

It must be pointed out, however, that one does not get in this way the original BSSN system. There is actually one difference in the lower order terms: only the principal parts are equivalent. The difference is (apart from the linear damping term) in the term of the form

$$+\frac{n}{2}\, Z^k \alpha_k\, \gamma_{ij} \tag{3.119}$$

in the evolution equation (3.110), which is missing in the original BSSN system [17]. This lower order term is needed for consistency with the general covariant equations (3.69).

We have seen how both the Bona–Massó and BSSN systems can be obtained from the more general Z4 formalism. The equivalence transformation (3.106) plays the crucial role because suppressing the Θ field (3.108) produces a sort of symmetry breaking: different values of the parameter n will lead to evolution systems that can no longer be transformed into one another once the set of dynamical fields is reduced by the vanishing of Θ.

We will extend this idea to the remaining supplementary quantity Z_i in the next chapter. Let us just note here that setting $Z_i = 0$ in the Z3 evolution system (3.109), (3.110), and (3.111) leads to a one-parameter family of non-equivalent free-evolution ADM systems. We can easily identify the $n = 0$ case as the 'Ricci evolution system' (2.60) and the $n = 1$ case as the 'Einstein evolution system' (2.63) that we introduced as separate options in the previous chapter.

References

1. C. Bona, J. Massó, E. Seidel and J. Stela, Phys. Rev. Lett. **75**, 600 (1995).
2. H. O. Kreiss and J. Lorentz, *Initial-Boundary Problems and the Navier-Stokes Equations*, Academic Press, New York (1989).
3. M. Alcubierre et al., Class. Quantum. Grav. **21**, 589 (2004).
4. W. H. Press, B. P. Flannery, S. A. Teukolsky and W. T. Veterling, *Numerical Recipes*, Cambridge University Press, Cambridge (1989).
5. L. Smarr and J. W. York, Phys. Rev. **D17**, 1945 (1978).
6. L. Smarr and J. W. York, Phys. Rev. **D17**, 2529 (1978).
7. J. Centrella, Phys. Rev. **D21**, 2776 (1980).
8. T. De Donder, *La Gravifique Einstenienne*, Gauthier-Villars, Paris (1921).
9. T. De Donder, *The Mathematical Theory of Relativity*, Massachusetts Institute of Technology, Cambridge (1927).
10. K. Lanczos, Ann. Phys. **13**, 621 (1922).
11. K. Lanczos, Z. Phys. **23**, 537 (1923).
12. V. A. Fock, *The Theory of Space, Time and Gravitation*, Pergamon, London (1959).
13. C. Bona and J. Massó, Phys. Rev. Lett. **68**, 1097 (1992).
14. T. Nakamura, K. Oohara and Y. Kojima, Prog. Theor. Phys. Suppl. **90**, 1 (1987).
15. T. Nakamura and K. Oohara, *Frontiers in Numerical Relativity*, ed. by C. R. Evans, L. S. Finn and D. Hobill, Cambridge University Press, Cambridge (1989).
16. M. Shibata and T. Nakamura, Phys. Rev. **D52**, 5428 (1995).
17. T. W. Baumgarte and S. L. Shapiro, Phys. Rev. **D59**, 024007 (1999).
18. A. Arbona, C. Bona, J. Massó and J. Stela, Phys. Rev. **D60**, 104014 (1999).
19. M. Alcubierre et al., Phys. Rev. **D64**, 061501 (2001).
20. M. Alcubierre et al., Phys. Rev. Lett. **87**, 271103 (2001).
21. M. Alcubierre et al., Phys. Rev. **D67**, 084023 (2003).
22. M. Campanelli, C. O. Lousto, P. Marronetti and Y. Zlochower, Phys. Rev. Lett. **96**,111101 (2006).
23. J. G. Baker et al., Phys. Rev. Lett. **96**, 111102 (2006).
24. C. Bona, T. Ledvinka and C. Palenzuela, Phys. Rev. **D66**, 084013 (2002).
25. C. Bona, T. Ledvinka, C. Palenzuela and M. Žáček, Phys. Rev. **D67**, 104005 (2003).
26. C. Bona, T. Ledvinka, C. Palenzuela and M. Žáček, Phys. Rev. **D69**, 064036 (2004).
27. F. Pretorius, Phys. Rev. Lett. **95**, 121101 (2005).
28. O. Brodbeck, S. Frittelli, P. Huebner and O. A. Reula, J. Math. Phys. **40**, 909 (1999).
29. C. Gundlach, G. Calabrese, I. Hinder and J. M. Martín-García, Class. Quantum Grav. **22**, 3767 (2005).
30. R. Arnowit, S. Deser and C. W. Misner. *Gravitation: An Introduction to Current Research*, ed. by L. Witten, Wiley, New York (1962). gr-qc/0405109.

Chapter 4
First-Order Hyperbolic Systems

From the mathematical point of view, the mixed-type systems (first order in time, second order in space) that we have considered in the previous chapter are associated with the parabolic type of equations. The prototype could be the Navier–Stokes equation of fluid dynamics, where second-order space derivatives appear in the viscosity terms. Parabolic equations are not the ones usually associated with causal propagation phenomena, where a finite propagation speed can be derived in a natural way from the governing equations.

Causal propagation is more easily described instead by systems of equations of hyperbolic type. The prototype is either the (second-order) wave equation, or the (first-order) Euler equations of fluid dynamics, where viscosity terms are not taken into account. The fact that computational fluid dynamics (CFD) deals mainly with hyperbolic first-order systems has stimulated the research on these systems, leading to interesting developments in applied mathematics, both at the theoretical and at the computational level. In order to take advantage of these advanced developments, it is convenient to express the mixed-type numerical relativity systems in a purely first-order form.

4.1 First-order versions of second-order systems

4.1.1 Introducing extra first-order quantities

A first-order version of the Z4 evolution system (3.80), (3.81), (3.82), and (3.83) can be obtained in the standard way by considering the first space derivatives:

$$A_k \;\equiv\; \partial_k\, ln\alpha, \quad B_k{}^i \;\equiv\; \partial_k\, \beta^i, \quad D_{kij} \;\equiv\; \frac{1}{2}\, \partial_k \gamma_{ij} \tag{4.1}$$

Bona, C. et al.: *First-Order Hyperbolic Systems.* Lect. Notes Phys. **783**, 79–108 (2009)
DOI 10.1007/978-3-642-01164-1_4

as independent dynamical quantities. This means that we must provide evolution equations for the new quantities (4.1). We will consider in this first section just normal coordinates (zero shift) for simplicity. The extra complications arising in the shift case will be addressed later in this chapter.

The standard way is just to realize that (4.1) are partial space derivatives of the metric components, so that the time derivatives of (4.1) will be mixed second derivatives of these coefficients. By reversing the order of space and time derivatives, one gets (normal coordinates)

$$\partial_t A_k \; + \; \partial_k[\, \alpha\, Q \,] = 0\,, \tag{4.2}$$

$$\partial_t D_{kij} \; + \; \partial_k[\, \alpha\, K_{ij} \,] = 0\,, \tag{4.3}$$

where we have introduced the time derivative quantity

$$Q \;\equiv -1/\alpha\, \partial_t\, ln\alpha\,, \tag{4.4}$$

which will of course depend on the time slicing condition. The choice (3.89) would correspond to

$$Q = f\,(tr\, K - m\, \Theta)\,. \tag{4.5}$$

Note that the new quantities A_k behave like the components of a 3+1 vector under general coordinate transformation, but the components D_{kij} do not transform in a covariant way. The full set of dynamical fields can be given by

$$\mathbf{u} \;=\; \{\alpha,\; \gamma_{ij},\; K_{ij},\; A_k,\; D_{kij},\; \Theta,\; Z_k\} \tag{4.6}$$

(38 independent fields).

Note also that the new quantities must be computed now through their evolution equations (4.2) and (4.3). The original definitions (4.1) are now considered rather as constraints, namely

$$\mathcal{A}_k \equiv A_k - \partial_k\, ln\alpha = 0 \tag{4.7}$$

$$\mathcal{D}_{kij} \equiv D_{kij} - \frac{1}{2}\, \partial_k \gamma_{ij} = 0\,. \tag{4.8}$$

These new first-order constraints are first integrals of the evolution equations (4.2) and (4.3), so that it is enough to enforce them on the initial data. It follows that the first-order versions will have a larger set of constraints than the original second-order systems. This is a complication, from both the theoretical and the computational point of view, as we will see in what follows.

Let us just remark that we can mimic here with the first-order constraints (4.7) and (4.8) the strategy adopted with the energy–momentum constraints in the previous chapter. Some ‘ad hoc’ damping terms can be added, devised to diminish first-order constraints violations, namely

$$\partial_t A_k \; + \; \partial_k[\, \alpha Q \,] = -\eta \, \mathcal{A}_k \tag{4.9}$$

$$\partial_t D_{kij} \; + \; \partial_k[\, \alpha K_{ij} \,] = -\eta \, \mathcal{D}_{kij} \,, \tag{4.10}$$

where the new damping coefficient η must be positive, as we will see below.

These damping coefficients can be assumed either to be constant or to include an α factor, depending whether we want the damping to act in coordinate time or in proper time, respectively.

4.1.2 Ordering ambiguities

The first complication related with the first-order constraints is the ordering ambiguities of second space derivatives. We can combine the space derivatives of the first equation in (4.1) to get an ordering constraint for A_k, namely

$$\mathcal{C}_{ij} \;\equiv\; \partial_i \, \mathcal{A}_j - \partial_j \, \mathcal{A}_i = \partial_i \, A_j - \partial_j \, A_i \;=\; 0 \,. \tag{4.11}$$

The fact that the ordering constraint (4.11) does not hold identically produces an ambiguity in the ordering of the second derivatives of the lapse, appearing in the evolution equation (3.81) for K_{ij} . This ordering ambiguity can be easily solved by taking the symmetric combination

$$\nabla_i \, d_j \, \alpha \;=\; \frac{1}{2}[\, \nabla_i(\alpha \, A_j) + \nabla_j(\alpha \, A_i) \,] \,, \tag{4.12}$$

which is the only one that preserves the symmetric character of K_{ij}.

We can get in the same way an ordering constraint for D_{kij}, namely

$$\mathcal{C}_{rsij} \;\equiv\; \partial_r \, \mathcal{D}_{sij} - \partial_s \, \mathcal{D}_{rij} = \partial_r \, D_{sij} - \partial_s \, D_{rij} \;=\; 0 \,. \tag{4.13}$$

When this ordering constraint does not hold identically, this produces again an ordering ambiguity in (3.81), where second space derivatives of the metric appear through the Ricci tensor R_{ij}.

But this time the ambiguity cannot be resolved by invoking the symmetric character of K_{ij}. On one side, we can get the standard Ricci decomposition

$${}^{(+)}R_{ij} \;=\; \partial_k \, \Gamma^k{}_{ij} - \partial_i \, \Gamma^k{}_{kj} + \Gamma^r{}_{rk}\Gamma^k{}_{ij} - \Gamma^k{}_{ri}\Gamma^r{}_{kj} \,, \tag{4.14}$$

where Γ_{kij} stands now for

$$\Gamma_{kij} \;\equiv\; D_{ijk} + D_{jik} - D_{kij}. \tag{4.15}$$

On the other hand, we can get the De Donder–Fock [1–5] decomposition

$$\begin{aligned} {}^{(-)}R_{ij} &= -\partial_k \, D^k{}_{ij} + \partial_{(i} \, \Gamma_{j)k}{}^k - 2D_r{}^{rk} D_{kij} \\ &\quad + 4D^{rs}{}_i D_{rsj} - \Gamma_{irs}\Gamma_j{}^{rs} - \Gamma_{rij}\Gamma^{rk}{}_k \end{aligned} \tag{4.16}$$

which is the preferred one in both the Bona–Massó and the BSSN systems, as discussed in Sect. 3.3.

There is no fundamental reason to prefer either the (4.14) or the (4.16) ordering. This is why we will consider an arbitrary combination of both cases, namely

$$R_{ij} = \frac{1+\zeta}{2}\, {}^{(+)}R_{ij} + \frac{1-\zeta}{2}\, {}^{(-)}R_{ij}\,, \tag{4.17}$$

where we have introduced the ordering parameter ζ so that the choice $\zeta = +1$ corresponds to the standard ordering (4.14), whereas the opposite choice $\zeta = -1$ corresponds to the alternative one (4.16). The intermediate value $\zeta = 0$ corresponds to the symmetrization of all second space derivatives.

Let us note at this point that the time evolution of the ordering constraints can be easily computed if we allow for the first-order equations (4.9) and (4.10), namely

$$\partial_t\, \mathcal{C}_{ij} = -\eta\, \mathcal{C}_{ij} \qquad \partial_t\, \mathcal{C}_{rsij} = -\eta\, \mathcal{C}_{rsij} \tag{4.18}$$

so we can see the role of the first-order damping coefficient, which will drive the ordering constraints toward zero (provided that $\eta > 0$).

4.1.3 First-order Z4 system (normal coordinates)

Now we are in a position to write down the first-order version of the Z4 evolution system (3.80), (3.81), (3.82), and (3.83) (normal coordinates)

$$\partial_t\, \alpha = -\alpha^2\, Q \tag{4.19}$$

$$\partial_t\, \gamma_{ij} = -2\, \alpha\, K_{ij} \tag{4.20}$$

$$\begin{aligned}\partial_t\, K_{ij} = & -\nabla_i \alpha_j + \alpha\, [\, R_{ij} + \nabla_i Z_j + \nabla_j Z_i \\ & -2\, K^2_{\ ij} + (trK - 2\Theta)\, K_{ij} - \kappa_1 (1+\kappa_2)\, \Theta\, \gamma_{ij}\,] \\ & -8\pi\alpha\, [\, S_{ij} - \frac{1}{2}\, (trS - \tau)\, \gamma_{ij}\,]\end{aligned} \tag{4.21}$$

$$\partial_t A_k = -\; \partial_k [\; \alpha\, Q - \eta\; ln\alpha\;] - \eta\, A_k \tag{4.22}$$

$$\partial_t D_{kij} = -\; \partial_k [\; \alpha\, K_{ij} - \eta\, \gamma_{ij}\;] - \eta\, D_{kij} \tag{4.23}$$

$$\begin{aligned}\partial_t\, \Theta = & \frac{\alpha}{2}\, [\, tr\, R + 2\, \nabla_k Z^k + (trK - 2\Theta)\, trK - tr(K^2) \\ & -2\, Z^k\, A_k - \kappa_1 (2+\kappa_2)\; \Theta - 16\pi\, \tau\,]\end{aligned} \tag{4.24}$$

$$\begin{aligned}\partial_t\, Z_i = & \alpha\, [\, \nabla_j\, (K_i^{\ j} - \delta_i^j\, trK) + \partial_i\, \Theta \\ & -2\, K_i^{\ j}\, Z_j - \Theta\, A_i - \kappa_1 Z_i - 8\pi\, S_i\,]\end{aligned} \tag{4.25}$$

where the second derivative terms in (4.21) must be replaced by (4.12) and (4.17) and the quantity Q in (4.19) and (4.22) can be obtained from any algebraic condition of the form (4.5).

The propagation properties of a first-order system like (4.19), (4.20), (4.21), (4.22), (4.23), (4.24), and (4.25) are given by the principal part terms, that is, the ones containing partial derivatives. It is then interesting for further purposes to write down the principal part of (4.19), (4.20), (4.21), (4.22), (4.23), (4.24), and (4.25) in an explicit way

$$\partial_t\, \alpha = \cdots \tag{4.26}$$

$$\partial_t\, \gamma_{ij} = \cdots \tag{4.27}$$

$$\partial_t\, \Theta + \partial_k\, \left[\alpha\, (D^k - E^k - Z^k)\right] = \cdots \tag{4.28}$$

$$\partial_t\, Z_i + \partial_k\, \left[\alpha\, (\, \delta_i^k\, (trK - \Theta) - K^k{}_i\,)\right] = \cdots \tag{4.29}$$

$$\partial_t\, A_k + \partial_k\, \left[\alpha\, f\, (trK - m\,\Theta) - \eta\, ln\alpha\right] = \cdots \tag{4.30}$$

$$\partial_t\, D_{kij} + \partial_k\, \left[\alpha\, K_{ij} - \eta\,\gamma_{ij}\right] = \cdots \tag{4.31}$$

$$\partial_t\, K_{ij} + \partial_k\, \left[\alpha\, \lambda^k_{ij}\right] = \cdots, \tag{4.32}$$

where the dots stand for terms not containing derivatives and we have noted for short

$$D_i \equiv D_{ik}{}^k\,, \qquad E_i \equiv D_{ki}{}^k\,, \tag{4.33}$$

$$\lambda^k_{ij} \equiv D^k{}_{ij} + \frac{1}{2}\, \delta^k_i (A_j + D_j - 2E_j - 2Z_j) + \frac{1}{2}\, \delta^k_j (A_i + D_i - 2E_i - 2Z_i)$$
$$- \frac{1+\zeta}{2}\, (D_{ij}{}^k + D_{ji}{}^k - \delta^k_i E_j - \delta^k_j E_i)\,. \tag{4.34}$$

Note that there is no ordering ambiguity in the evolution equation (4.28) for Θ. This is because the trace of the ζ-dependent terms in (4.34) vanishes identically. One has the full set of 38 evolution equations for the 38 fields in (4.6), containing f and m as gauge parameters, ζ as ordering parameter, and the κ and η damping parameters for energy–momentum and first-order constraints, respectively.

4.1.4 Symmetry breaking: the KST system

We have obtained in the previous sections the first-order version (4.19), (4.20), (4.21), (4.22), (4.23), (4.24), and (4.25) of the Z4 system by assuming the standard evolution equations (4.22) and (4.23) for the first-order quantities (4.1). But this is not the only way of getting a first-order version of a second-order system. Kidder, Scheel, and Teukolsky obtained a first-order version of the original ADM system by combining the energy and momentum constraints with the first-order evolution equations (KST system [6]).

The original KST system assumed a direct power-law relationship between the lapse and the space volume element ('densitized' lapse). This system has been extended by Sarbach and Tiglio [7] in order to include the wide class of generalized harmonic conditions that we are considering here, in which the lapse is an independent quantity which must be computed from its evolution equation ('dynamical' lapse). In what follows we will use the term 'KST system' to refer also to this generalization.

We will recover here the KST system from the Z4 one by a symmetry breaking mechanism, along the lines sketched in Sect. 3.5.3. Let us consider as a starting point the principal part (4.26), (4.27), (4.28), (4.29), (4.30), (4.31), and (4.32), dropping here the damping terms for simplicity. We will follow a two-step 'symmetry breaking' process, namely

1. Recombining the dynamical fields K_{ij}, D_{kij} with Θ and Z_i in a linear way,

$$\tilde{K}_{ij} = K_{ij} - \frac{n}{2}\,\Theta\,\gamma_{ij}\,, \tag{4.35}$$

$$d_{kij} = 2\,D_{kij} + \eta\,\gamma_{k(i}Z_{j)} + \chi\,Z_k\,\gamma_{ij}\,, \tag{4.36}$$

   where we have used the notation of [6], replacing only their parameter γ by $-n/2$ for consistency with the definition (3.106) in Sect. 3.5.3.
2. Suppressing both θ and Z_i as dynamical fields, namely

$$\Theta = 0\,, \qquad Z_i = 0\,. \tag{4.37}$$

Note that the linear combinations (4.35) and (4.36) are generic in the sense that they are the most general linear combinations that preserve the tensor character of the dynamical fields under linear coordinate transformations (remember that the D's components are not covariant under general coordinate transformations).

In that way, the principal part (4.27), (4.28), (4.29), (4.30), (4.31), and (4.32) becomes

$$\partial_t\,\alpha = \cdots \tag{4.38}$$

$$\partial_t\,\gamma_{ij} = \cdots \tag{4.39}$$

$$\partial_t\,A_k + \partial_k\,[\,\alpha\,f\,tr\,\tilde{K}\,] = 0 \tag{4.40}$$

$$\begin{aligned}\partial_t\,d_{kij} + \partial_r[\,\alpha\,\{2\,\delta^r_k\,\tilde{K}_{ij} - \chi\,(\tilde{K}^r_k - \delta^r_k\,tr\,\tilde{K})\,\gamma_{ij}\\ +\eta\,\gamma_{k(i}(\tilde{K}^r{}_{j)} - \delta^r{}_{j)}tr\,\tilde{K})\}\,] = \cdots\end{aligned} \tag{4.41}$$

$$\partial_t\,\tilde{K}_{ij} + \partial_k\,[\,\alpha\,\lambda^k_{ij}\,] = \cdots \tag{4.42}$$

for the reduced set of 34 dynamical fields

$$\mathbf{u} = \{\alpha,\,\gamma_{ij},\,\tilde{K}_{ij},\,A_k,\,d_{kij}\,\}\,, \tag{4.43}$$

where λ^k_{ij} stands now for

$$
\begin{aligned}
2\,\lambda^k{}_{ij} \equiv d^k{}_{ij} &- \frac{n}{4}\,(d_{kr}{}^r - d_r{}^{rk})\,\gamma_{ij} \\
&+ \frac{1+\zeta}{2}\,(d_{ij}{}^k + d_{ji}{}^k) - \frac{1-\zeta}{2}\,(\delta^k_i\, d_{rj}{}^r + \delta^k_j\, d_{ri}{}^r) \\
&+ \delta^k_j\,(A_i + \frac{1}{2}\,d_{ir}{}^r) + \delta^k_i\,(A_j + \frac{1}{2}\,d_{jr}{}^r)\,. \qquad (4.44)
\end{aligned}
$$

This provides the 'dynamical lapse' version [7] of the KST evolution system. In order to recover the original 'densitized lapse' version, one must in addition integrate explicitly the dynamical relationship (4.5) between the lapse and the volume element (remember that now $\Theta = 0$). It can be easily done in the case

$$
f \;=\; 2\,\sigma \;=\; \text{constant}, \qquad (4.45)
$$

namely

$$
\partial_t\,(\,\alpha\gamma^{-\sigma}) \;=\; 0, \qquad (4.46)
$$

so that the value of α can be defined in terms of the space volume element $\sqrt{\gamma}$ for every initial condition. The same thing can be done with their space derivatives, so that we can take

$$
A_i \;\equiv\; \sigma\, d_{ir}{}^r\,, \qquad (4.47)
$$

and the set of dynamical fields is then further reduced to

$$
\mathbf{u} \;=\; \{\,\gamma_{ij},\; K_{ij},\; d_{kij}\,\}. \qquad (4.48)
$$

The principal part of the evolution system is then given by (we suppress the tildes over the K_{ij})

$$
\begin{aligned}
&\partial_t\;\gamma_{ij} = \cdots \qquad (4.49)\\
&\partial_t\; d_{kij} + \partial_r[\;\alpha\{\,2\,\delta^r_k\,K_{ij} - \chi\,(K_k{}^r - \delta^r_k\,tr\,K)\,\gamma_{ij} \\
&\qquad\qquad + \eta\;\gamma_{k(i}(K^r{}_{j)} - \delta^r_{j)}tr\,K)\,\}\;] = \cdots \qquad (4.50)\\
&\partial_t\;K_{ij} + \partial_k\;[\;\alpha\;\lambda^k_{ij}\;] = \cdots \qquad (4.51)
\end{aligned}
$$

with

$$
\begin{aligned}
2\,\lambda^k{}_{ij} \equiv d^k{}_{ij} &- \frac{n}{4}\,(d_{kr}{}^r - d_r{}^{rk})\,\gamma_{ij} + \frac{1+2\sigma}{2}\,(\delta^k_i\,d_{j\,r}{}^r + \delta^k_j\,d_{ir}{}^r) \\
&- \frac{1-\zeta}{2}\,(\delta^k_i\,d^r{}_{rj} + \delta^k_j\,d^r{}_{ri}) + \frac{1+\zeta}{2}\,(d_{ij}{}^k + d_{j\,i}{}^k), \qquad (4.52)
\end{aligned}
$$

which corresponds precisely to (the principal part of) the original KST system [6].

Note that we have lost in the process the second gauge parameter m, and the first one f has been replaced by σ. On the other hand, we have kept the ordering parameter ζ and we have got one extra energy-constraint parameter n and two extra momentum constraint parameters χ and η.

4.2 Hyperbolic systems

We have seen in Sect. 3.1.3 how to study the propagation properties of the field equations by means of the plane-wave analysis. This is a very general, physically sound method which provides consistent results, independent of the way we manage to write down the equations.

Now we will see still another method: the hyperbolicity analysis, which is currently used in the mathematical literature for first-order systems [8] and can then be applied to the first-order form of our evolution equations. Contrary to the plane-wave analysis, the hyperbolicity analysis takes into account just the principal part of the system: the terms containing the higher derivatives for every dynamical field. This prevents a consistent treatment of the linear damping terms, which will be switched off in this section. We will see then that the quasilinear nature of Einstein's field equations leads to a full coincidence of the results obtained by these two methods.

4.2.1 Weak and strong hyperbolicity

Hyperbolic first-order systems have been proposed for numerical relativity applications since the seminal work of Y. Choquet-Bruhat and T. Ruggeri [9]. In all of them, the original ADM system [10] is modified by using the constraints in one or the other way. This includes the Bona–Massó system [11, 12], the KST one [6, 7], the generalized harmonic one [13, 14] and some others [15–19], even taking additional derivatives in some cases [20, 21] (but see [22, 23] for a completely different approach). The first-order version of the BSSN system (NOR system [24]) has been also analyzed from this point of view.

We will consider here a generic first-order system, although the specific developments will be carried out for the first-order version (4.19), (4.20), (4.21), (4.22), (4.23), (4.24), and (4.25) of the Z4 system [25]. For the purposes of our analysis, we will need to deal only with the principal part, that is, the one containing first derivatives of the basic fields $\mathbf{u}$. We will write then our first-order system in the form

$$\partial_t\, \mathbf{u} + \mathbf{A}^k\, \partial_k\, \mathbf{u} = 0\,, \tag{4.53}$$

where $\mathbf{A}^k$ is the 'characteristic' matrix along the k axis.

Let us consider now a generic space direction, given by the unit vector $\boldsymbol{n}$. We will study the following eigenvalue problem

$$(\mathbf{A}^n - v\,\mathbf{I})\,\mathbf{u} = 0\,, \tag{4.54}$$

where the matrix $\mathbf{A}^n = n_k \mathbf{A}^k$ will be called again the characteristic matrix along the direction $\boldsymbol{n}$. Their eigenvalues v will be the corresponding 'characteristic speeds.'

We will say that the first-order system (4.53) is

- **Strongly hyperbolic** if, for every direction $\boldsymbol{n}$, all the characteristic speeds are real and the characteristic matrix can be put into full diagonal form (there is a complete set of eigenvectors).
- **Weakly hyperbolic** if, for every direction $\boldsymbol{n}$, all the characteristic speeds are real but, at least for some direction, the characteristic matrix cannot be fully diagonalized.

These definitions coincide with the strong and weak pseudo-hyperbolicity, respectively, that were introduced in Sect. 3.1.3, when applied to the homogeneous first-order system (4.53). By this we mean that, in the linear approximation, the dynamics described by (the principal part of) a strongly hyperbolic system is just linear wave propagation along the selected direction, with the characteristic speed as the wave speed.

Hyperbolic systems can be shown, with some additional smoothness assumptions [8], to have a well-posed initial-value problem. This implies the stability of the solutions at the continuum level, which is a necessary condition for the stability of numerical solutions. We have seen in Sect. 3.2 that strong hyperbolicity is actually required to get stability at the numerical level. We will see in the following chapter how this property is also important in order to apply some advanced numerical methods from computational fluid dynamics.

Before going further in that direction, let us solve the eigenvalue problem for the first-order version (4.19), (4.20), (4.21), (4.22), (4.23), (4.24), and (4.25) of the Z4 system. Let us note first that the principal terms of the evolution equations (4.26) and (4.27) for the metric coefficients are trivial, so that we can restrict ourselves to the reduced set of 31 dynamical fields

$$\mathbf{u} \;=\; \{K_{ij},\ A_k,\ D_{kij},\ \Theta,\ Z_k\}\,, \tag{4.55}$$

which can be considered as evolving linearly in a non-homogeneous background provided by the metric coefficients α, γ_{ij}. Let us also introduce the first-order version of the Bona–Massó quantity (3.115)

$$V_k = D_k - E_k - Z_k \tag{4.56}$$

as an auxiliary variable which will prove very useful for the analysis.

The spectral decomposition of the characteristic matrix $\mathbf{A}^n$ provides the following list of eigenfields:

- **Standing eigenfields** (zero characteristic speed):

$$A_{\perp},\ D_{\perp ij},\ A_k - f D_k + f\, m V_k \tag{4.57}$$

(17 independent fields), where the symbol $\perp$ replacing an index means the projection orthogonal to n_i, for instance

$$D_{\perp ij} \equiv (\delta_k^{\ r} - n_k n^r)\, D_{rij}\,. \tag{4.58}$$

- **Light-cone eigenfields** (local characteristic speed $\pm\alpha$):

$$L_{ij}^{\pm} \equiv [\, K_{ij} - n_i n_j\, trK\,] \pm [\, \lambda^n{}_{ij} - n_i n_j\ tr\lambda^n\,] \tag{4.59}$$
$$E^{\pm} \equiv \theta \pm V^n \tag{4.60}$$

(12 independent fields), where the symbol n replacing the index means the contraction with n_i, for instance

$$\lambda^n_{ij} \ \equiv\ n_k \lambda^k_{ij}\,. \tag{4.61}$$

- **Gauge eigenfields** (characteristic speed $\pm\alpha\sqrt{f}$). In the generic, non-degenerate case ($f \neq 1$), we get

$$G^{\pm} \equiv \sqrt{f}\,[\ tr\, K - \mu\, \Theta\ ] \pm [\ A^n + (2-\mu)\ V^n\ ]\,, \tag{4.62}$$

where we have written for short

$$\mu = \frac{f\, m - 2}{f - 1}\,. \tag{4.63}$$

In the degenerate case $f = 1$, one must have $m = 2$, as discussed below, and the extra degeneracy allows any value of the parameter μ in the combination (4.62). The corresponding eigenfields can be chosen to be, for instance,

$$[\ tr\, K\ ] \pm [\ A^n + 2\ V^n\ ]\,. \tag{4.64}$$

From the list ((4.57), (4.58), (4.59), (4.60), (4.61), and (4.62)) of eigenvectors and their corresponding eigenvalues, we can easily conclude that

- All the characteristic speeds are real (the system is at least weakly hyperbolic) if and only if $f \geq 0$.
- In the case $f = 0$, the two components of the gauge pair $G^{\pm}$ are not independent, so that the total number of independent eigenfields is 30 instead of the 31 ones required for strong hyperbolicity.
- The case $f = 1$ (harmonic case) is special:

- If $m \neq 2$, then the gauge pair $G^{\pm}$ cannot be fully decoupled from the pair $E^{\pm}$, so that one has only 29 independent eigenfields.
- If $m = 2$, then these pairs can be decoupled in many ways, due to the degeneracy of the gauge and light eigenfields. The pair (4.64) is just one of such choices. One recovers then the full set of 31 independent eigenfields (strong hyperbolicity).

• The first-order Z4 system described by (4.19), (4.20), (4.21), (4.22), (4.23), (4.24), and (4.25) is strongly hyperbolic in all the remaining cases ($f > 0$, $f \neq 1$).

4.2.2 1D Energy estimates

In order to gain some further insight, let us focus for the moment in the one-dimensional (1D) case. There is only one space direction and then just one characteristic matrix. Let us assume strong hyperbolicity, so that the corresponding left eigenvectors actually provide a complete basis of the dynamical fields space,

$$\mathbf{w} \; = \; \{ w_r \} \qquad r, s = 1, \ldots, n \,, \tag{4.65}$$

where n is the number of independent dynamical fields.

We have found these eigenfields by inspection in the preceding subsection, due to the simple structure of the Z4 first-order system in normal coordinates. This is not always so simple, as we will see for instance in the magnetohydrodynamics (MHD) case. A more systematic way is to relate the eigenfields $\mathbf{w}$ with the original fields $\mathbf{u}$ by

$$w_r = \sum_s L_{rs} \, u_s \,, \tag{4.66}$$

where the coefficients

$$(\, L_{r1} \,, \, \ldots, \, L_{rn} \,) \qquad r = 1, \ldots, n \tag{4.67}$$

are the left eigenvectors of the characteristic matrix.

In the linear approximation, we can assume that the diagonalization matrix L is constant, so that the original system (4.53) can be written in the basis $\mathbf{w}$ as a set of uncoupled advection equations, namely

$$\partial_t \, w_r + \lambda_r \, \partial_x \, w_r = 0 \qquad r = 1, \ldots, n \,. \tag{4.68}$$

It follows that, in the linear regime, the dynamics can be described as a superposition of simple waves, propagating with the characteristic speeds, in full concordance with the plane-wave analysis results.

Remember that the full diagonalization of the characteristic matrix is not possible in the weakly hyperbolic case. A prototype example could be the two-component system

$$\partial_t \begin{pmatrix} u \\ v \end{pmatrix} + \begin{pmatrix} \lambda & a \\ 0 & \lambda \end{pmatrix} \partial_x \begin{pmatrix} u \\ v \end{pmatrix} = 0\,. \tag{4.69}$$

There is just one independent eigenfield (v), and the evolution of the remaining component u contains a linearly growing mode, namely

$$u(x,t) = f(x - \lambda t) - a\,t\,v'(x - \lambda t)\,. \tag{4.70}$$

These polynomial modes (linear in this simple example) are easily detected by the robust stability test, as described in Sect. 3.2.

Coming back to the strong hyperbolicity case, we can introduce a positive-definite quadratic form in the space of dynamical fields:

$$E = \sum_r (w_r)^2 \tag{4.71}$$

(energy metric). We will consider the 'energy estimate' $\overline{E}$, obtained as the integrated value of E over the computational domain

$$\overline{E} = \int_a^b E\;dx = \sum_r \int_a^b (w_r)^2\;dx\,. \tag{4.72}$$

Allowing for the uncoupled evolution equations (4.68), we can obtain the time evolution of this estimate as

$$\partial_t\,\overline{E} = -2\,\sum_r\;\lambda_r \int_a^b w_r\;(\partial_x w_r)\;dx = -\left[\sum_r \lambda_r (w_r)^2\right]_a^b\,. \tag{4.73}$$

This means that, in the linear approximation, the 'energy estimate' $\overline{E}$ is conserved, apart from boundary term contributions. We will assume for the moment that the boundary is far from the dynamical region (or, alternatively, periodic boundary conditions) so that the boundary term contribution can be ignored here: it will be discussed properly later in this chapter.

The conservation of the energy estimate provides a direct way to show that the system under consideration is well-posed. If every dynamical field at every grid point contributes to the energy estimate $\overline{E}$ (which is positive definite, so that there can no be cancellations of any kind), this means that the value of any specific dynamical field cannot grow without bound. This was already implicit in the evolution equations (4.68), but the energy estimate allows to focus the analysis in a single quantity, the energy metric, which is much more convenient than dealing with the full characteristic decomposition.

Note, however, that the energy metric (4.71) is far from being unique. To begin with, the characteristic eigenfields w_r are just defined up to an arbitrary factor. Moreover, when some characteristic speeds coincide, the linear degeneracy of the corresponding eigenspace produces non-equivalent quadratic forms which could be used as alternative starting points, leading to different energy estimates. There are also other more general choices, as we will see in what follows.

4.2.3 Symmetric-hyperbolic systems

Let us go back to the multidimensional case, to the linear system

$$\partial_t\ \mathbf{u} + \mathbf{A}^k\,\partial_k\,\mathbf{u} = 0\,, \tag{4.74}$$

where we will assume again that the characteristic matrices $\mathbf{A}^k$ have constant coefficients in the linear approximation. This system will be said to be symmetric-hyperbolic if there exists a basis of dynamical fields $\mathbf{u}$ such that, for every space direction $\boldsymbol{n}$, the corresponding characteristic matrix is symmetric, that is,

$$n_k {A^k}_{rs} = n_k {A^k}_{sr} \qquad \forall \boldsymbol{n}\,. \tag{4.75}$$

It is clear that any symmetric-hyperbolic system must be also strongly hyperbolic: every real symmetric matrix can be fully diagonalized and all the corresponding eigenvalues are real. The converse is not true: there are as many characteristic matrices as space directions and the requirement that all of them must be symmetric when expressed in a given basis is not trivial.

Let us restrict ourselves here to symmetric-hyperbolic systems. Let $\mathbf{u}$ be one basis in which the symmetry requirement (4.75) is fulfilled. We will then define the energy metric to be

$$E = \sum_r (u_r)^2\,. \tag{4.76}$$

Now some comments are in order:

- In the 1D case, using the basis of eigenfields $\mathbf{w}$ ensures that the (only) characteristic matrix is in diagonal form (that is symmetric, of course). This means that symmetric hyperbolicity is equivalent to strong hyperbolicity in the 1D case.
- In the 3D case, there is no common basis of eigenvectors for the characteristic matrices along different directions (unless all these matrices commute). This means that the basis fields $\mathbf{u}$ are not eigenvectors in the generic case.
- This suggests a supplementary freedom in choosing the basis vectors for the energy metric: any combination of the form

$$u_r = \sum_s R_{rs} w_s \,, \tag{4.77}$$

where $\mathbf{R}$ is any orthogonal matrix, will work, as far as the symmetry of a matrix is preserved under orthogonal transformations. Note that the resulting fields u_r will not be eigenvectors in the generic case.

Now we can proceed along the same lines as in the 1D case. We will consider again the energy estimate, obtained by integrating E over the computational domain

$$\overline{E} = \int E \; dV = \sum_r \int (u_r)^2 \; dV \,. \tag{4.78}$$

Its time variation is given by (we are using the symmetry requirement here)

$$\partial_t \, \overline{E} = -2 \, \sum_{rs} {A^k}_{rs} \int u_r \, \partial_k u_s \; dV = - \oint \sum_{rs} \; n_k {A^k}_{rs} \, u_r u_s \; dS \,, \tag{4.79}$$

where the surface integral is over the boundary of the computational domain. Note that we need only one characteristic matrix at a time in order to compute the last integral: the one corresponding to the normal direction $\boldsymbol{n}$ to the given boundary surface, that is,

$$\mathbf{A}^n = n_k \mathbf{A}^k \,. \tag{4.80}$$

This means that energy estimates can be useful for devising boundary conditions for multidimensional symmetric-hyperbolic systems in the same way as in the 1D case, as we will discuss later in this chapter.

Let us recall here that symmetric hyperbolicity implies strong hyperbolicity, but not vice versa. Symmetric hyperbolicity is not required in order to get a well-posed system in the context of the pure initial data problem. The advantage of this stronger requirement is rather that the knowledge of the energy estimates can provide useful clues for devising additional conditions, for instance in the context of the mixed initial-boundary problem. But note also that this could be in conflict with other requirements, like the current prescriptions for singularity-avoidant coordinates. We will regard then symmetric hyperbolicity just as a bonus, not such a strict requirement as strong hyperbolicity.

A simple example of a symmetric-hyperbolic system in normal coordinates is provided by the first-order version (4.19), (4.20), (4.21), (4.22), (4.23), (4.24), and (4.25) of the Z4 system in the harmonic slicing case ($f = 1, \; m = 2$), for the choice $\zeta = -1$ of the ordering parameter [26]. The energy metric can be chosen to be

$$E = K_{ij} K^{ij} + \lambda_{kij} \lambda^{kij} + \Theta^2 + V_k V^k \,, \tag{4.81}$$

but a 'constant' term of the form

$$(A_k - D_k + 2V_k)\,(A^k - D^k + 2V^k)\,, \tag{4.82}$$

containing just fields with zero characteristic speed, could also be added. Note that, as discussed before, the quadratic form (4.81) is built with simple combinations of the basic dynamical fields, without requiring the explicit use of any characteristic basis.

4.3 Generic space coordinates

4.3.1 First-order fields

Let us now consider the general case, beyond the normal coordinates prescription. This amounts to considering the generic shift case, so that the shift-related first-order constraint in (4.1), namely

$$\mathcal{B}_k{}^i \equiv B_k{}^i - \partial_k\,\beta^i = 0\,, \tag{4.83}$$

will play an outstanding role.

The first decision to take is how to translate the evolution equations for the metric components. Let us consider for instance (3.80), namely

$$(\partial_t - \mathcal{L}_\beta)\;\gamma_{ij} = -2\,\alpha\,K_{ij}\,. \tag{4.84}$$

At first sight, one could be tempted to write it in the most direct way, as (3.80), namely

$$(\partial_t - \beta^k\,\partial_k)\;\gamma_{ij} - \gamma_{ik}\,\partial_j\beta^k - \gamma_{kj}\,\partial_i\beta^k = -2\,\alpha\,K_{ij}\,. \tag{4.85}$$

This is not a good idea, because space derivatives of the metric coefficients (β^k and γ_{ij} in this case) would appear explicitly in the equations. In this way, we could no longer consider these coefficients as a sort of kinematical background, unaffected by space discretization errors, the first-order quantities $(K_{ij}\,,\; D_{kij}\,,\;\ldots)$ being the truly dynamical degrees of freedom. From this point of view, the rule of thumb should be using the constraints (4.7), (4.8), and (4.83) to replace systematically the metric space derivatives. The only exceptions could be those required by keeping the equations in flux-conservative form, as we will see later, or by the damping terms that we have seen before.

We will express then (4.84) in the more convenient form

$$\partial_t\;\gamma_{ij} = 2\,\beta^k D_{kij} + B_{ij} + B_{j\,i} - 2\,\alpha\,K_{ij}\,, \tag{4.86}$$

where we have noted

$$B_{ki} = \gamma_{ij} \, B_k{}^j \,. \tag{4.87}$$

The evolution equations for the remaining metric coefficients will be given then by

$$\partial_t \, \alpha = \alpha \, \beta^k A_k - \alpha^2 \, Q \,, \qquad \partial_t \, \beta^i = \beta^k \, B_k{}^i - \alpha \, Q^i \,, \tag{4.88}$$

where the quantities $(Q, \, Q^i)$ must be provided by the coordinate gauge prescription.

Note that replacing $B_k{}^i$ by the corresponding shift derivative in (4.88) would introduce a truly non-linear behavior, similar to that of the Euler (or Burgers) equations. As we will see in the next chapter, this can be a source of numerical problems. This was actually the reason why the initial first-order versions of the generalized harmonic formalism [13] were replaced by the most recent ones, like [14], where more general combinations of (4.84) and (4.86) are considered in order to obtain a better behaved system.

The same logic applies to the evolution equations (4.2) and (4.3) of the space-derivatives quantities. The most direct translation could be (we switch off the damping terms for the moment)

$$\partial_t \, A_k \;+\; \partial_k \, [\, -\beta^r A_r + \alpha \, Q \,] = 0 \,, \tag{4.89}$$

$$\partial_t \, B_k{}^i \;+\; \partial_k \, [\, -\beta^r B_r{}^i + \alpha \, Q^i \,] = 0 \,, \tag{4.90}$$

$$\partial_t \, D_{kij} \;+\; \partial_k \, [\, -\beta^r D_{rij} + \alpha \, Q_{ij} \,] = 0 \,, \tag{4.91}$$

where we have noted for short

$$Q_{ij} \equiv K_{ij} - \frac{1}{2\alpha} \, (B_{ij} + B_{j\,i}) \,. \tag{4.92}$$

But, again, this may not be a good idea. Note that the transverse derivative components in (4.89), (4.90), and (4.91), namely

$$A_\perp \,, \qquad B_\perp{}^i \,, \qquad D_{\perp ij} \,, \tag{4.93}$$

do propagate along the time lines (zero characteristic speed). This is a complication, when the main dynamical fields will propagate instead along light cones. Time lines can cross the local light cone in simulations involving superluminal regions and the resulting propagation speed degeneracy could prevent the full diagonalization of the characteristic matrix (weak hyperbolicity).

A safer alternative is using (4.83) plus the shift ordering constraint

$$\mathcal{C}_{rs}{}^i \equiv \partial_r \, \mathcal{B}_s{}^i - \partial_s \, \mathcal{B}_r{}^i = \partial_r \, B_s{}^i - \partial_s \, B_r{}^i = 0 \tag{4.94}$$

for transforming the evolution equations (4.89), (4.90), and (4.91) into

$$\partial_t A_k \,+\, \partial_l [-\beta^l \; A_k + \delta^l_k \, \alpha \, Q] = B_k{}^l \, A_l - trB \, A_k \tag{4.95}$$

$$\partial_t B_k{}^i \,+\, \partial_l [-\beta^l \; B_k{}^i + \delta^l_k \, \alpha \, Q^i] = B_k{}^l \, B_l{}^i - trB \, B_k{}^i \tag{4.96}$$

$$\partial_t D_{kij} \,+\, \partial_l [-\beta^l D_{kij} + \delta^l_k \, \alpha \, Q_{ij} \,] = B_k{}^l \, D_{lij} - trB \, D_{kij}. \tag{4.97}$$

In this way, the transverse derivatives components (4.93) will propagate along the normal lines (characteristic speed $-\beta$), which of course cannot cross the local light cone for a spacelike constant-time hypersurface.

Some remarks are in order at this point:

- We realize here that using the ordering constraints can actually affect the characteristic speeds and the hyperbolicity of the first-order evolution system. This is a feature of the shift case: as we have seen in Sect. 4.2.1, the ordering parameter ζ just affected the computation of the eigenvectors, not the eigenvalues. Only when demanding the stronger requirement of symmetric hyperbolicity a particular value ($\zeta = -1$) was preferred.
- These results can be actually confirmed by the plane-wave analysis, provided that we treat consistently ${B_k}^i$ as independent quantities. We should consider again a constant background (with a non-zero shift this time). The (quadratic) right-hand-side terms in (4.95), (4.96), and (4.97) will not be taken into account in the linear approximation. This can seem unphysical, but it is the right way for analyzing the behavior of our first-order system under generic perturbations arising from numerical truncation errors.

Let us finally analyze the resulting evolution equations for the first-order constraints. We will consider first the lapse-derivatives constraint (4.7). Allowing for (4.95), we get

$$\partial_t\, \mathcal{A}_k - \beta^r\,(\partial_r\, \mathcal{A}_k - \partial_k\, \mathcal{A}_r) = {\mathcal{B}_k}^r\, A_r - {\mathcal{B}_r}^r\, A_k\,. \tag{4.98}$$

The hyperbolicity of the subsidiary evolution equation (4.98) can be analyzed by displaying the normal and transverse components of the principal part along any space direction $\boldsymbol{n}$, namely

$$\partial_t\, \mathcal{A}_n - \beta^\perp\,(\partial_n\, \mathcal{A}_\perp) = 0 \tag{4.99}$$

$$\partial_t\, \mathcal{A}_\perp - \beta^n\,(\partial_n\, \mathcal{A}_\perp) = 0\,, \tag{4.100}$$

with eigenvalues $(0,\ -\beta^n)$, which is just weakly hyperbolic in the fully degenerate case, that is, for any space direction orthogonal to the shift vector.

This fact explains the importance of adding constraint damping terms to (4.95), namely

$$\partial_t A_k + \partial_l[-\beta^l A_k + \delta^l_k\,\alpha\, Q\,] = {B_k}^l\, A_l - trB\, A_k - \eta\, \mathcal{A}_k\,, \tag{4.101}$$

so that the damping term $-\eta\,\mathcal{A}_k$ will appear as a result in the subsidiary system also. The linearly growing constraint-violation modes arising from the degenerate coupling in (4.99) will be kept then under control by these (exponential) damping terms. The same argument applies *mutatis mutandis* to the remaining first-order constraints ${\mathcal{B}_k}^i$, $\mathcal{D}_{kij}$.

4.3.2 Generalized harmonic formulations

As a first example of strongly hyperbolic formalism with dynamical shift prescription, we will consider here the first-order versions of the generalized harmonic formalism (1.82), namely

$$\frac{1}{2}\, g^{\rho\sigma}\partial^2_{\rho\sigma}\; g_{\mu\nu} + \partial_{(\mu}H_{\nu)} - \Gamma^{\rho}_{\mu\nu}H_\rho \tag{4.102}$$
$$= g^{\rho\sigma}g^{\lambda\tau}\,[\,\partial_\lambda g_{\mu\rho}\;\partial_\tau g_{\nu\sigma} - \Gamma_{\mu\rho\lambda}\;\Gamma_{\nu\sigma\tau}\,] - 8\pi\,(T_{\mu\nu} - \frac{T}{2}\,g_{\mu\nu})\,.$$

Let us just remember that the coordinate gauge choice here amounts to the prescription of the gauge sources H_μ. Notice that we are not considering here any constraint-damping terms.

We will consider for simplicity the cases in which the principal part of the evolution system is just a set of scalar wave equations, that is,

$$g^{\rho\sigma}\partial^2_{\rho\sigma}\; g_{\mu\nu} = \cdots \tag{4.103}$$

This simplification can be obtained in many different ways:

- The gauge sources H_μ can be given by some kinematical prescription. This prescription can involve the metric coefficients, but not their first derivatives. For instance, in the standard harmonic case, we get $H_\mu = 0$.
- The gauge sources can be considered to be independent dynamical fields. Additional evolution equations must then be prescribed for these fields, which must be at least of second order in H_μ, so that the first-derivative terms in (4.102) do not belong to the principal part. For instance, we can take
$$\Box H_\mu = \cdots\,. \tag{4.104}$$
- A combination of the two mechanisms described above. We can take for instance [27]
$$H_i = 0\,, \qquad \Box H_n = \cdots\,, \tag{4.105}$$
where $H_n \equiv n^\mu H_\mu$ is the projection orthogonal to the constant time slices.

The simple structure of (4.103) suggests using here a simplified 3+1 decomposition, in which every single metric component is dealt with as an independent function. The first metric derivatives will be decomposed then in the following way:

$$D_{k\mu\nu} \equiv \partial_k\, g_{\mu\nu} \tag{4.106}$$
$$Q_{\mu\nu} \equiv n^\rho\partial_\rho\, g_{\mu\nu} = -\frac{1}{\alpha}(\partial_t - \beta^k\partial_k)\, g_{\mu\nu}\,, \tag{4.107}$$

where we have dropped some 1/2 factors, so that the space components D_{kij} do not coincide with the ones previously considered in the standard 3+1 framework. The second-order evolution equations (4.103) can then be translated into the first-order system

$$\partial_t \; g_{\mu\nu} \;= \beta^k D_{k\mu\nu} - \alpha\, Q_{\mu\nu} \qquad (4.108)$$

$$(\partial_t - \beta^k \partial_k)\; Q_{\mu\nu} + \alpha\, \partial_k(\gamma^{kj} D_{j\,\mu\nu}) = \cdots \qquad (4.109)$$

Considering just the principal part, we can assume here again that the first-derivatives dynamical fields ($Q_{\mu\nu}$, $D_{k\mu\nu}$) propagate in an inhomogeneous background defined by the metric coefficients $g_{\mu\nu}$.

Some evolution equations must now be prescribed for the space derivatives quantities $D_{k\mu\nu}$. For the reasons explained in the preceding subsection, we will take

$$\partial_t D_{k\mu\nu} + \partial_l[-\beta^l D_{k\mu\nu} + \delta^l_k\, \alpha\, Q_{\mu\nu}\,] = B_k{}^l\, D_{l\mu\nu} - trB\, D_{k\mu\nu}\,. \qquad (4.110)$$

In this way, the characteristic fields of the first-order harmonic system (4.109) and (4.110) associated with any given space direction n can be decomposed into two sets:

- **The transverse derivatives** $D_{\perp\mu\nu}$, propagating along the normal lines (characteristic speed $-\beta^n$).
- **The light-cone eigenfields**, given by the pairs

$$Q_{\mu\nu} \pm D_{n\mu\nu}, \qquad (4.111)$$

with characteristic speed $-\beta^n \pm \alpha$, respectively.

Here again, the index n means the projection along the selected space direction, whereas the index $\perp$ stands for the orthogonal projection. It is then evident that we have a complete set of eigenfields, so that the system (4.109) and (4.110) is strongly hyperbolic. Moreover, it is symmetric hyperbolic, with the quadratic form

$$E \equiv \sum_{\mu\nu}\; [\, Q_{\mu\nu}\, Q_{\mu\nu} + \gamma^{kr}\, D_{k\mu\nu}\, D_{r\mu\nu}\,] \qquad (4.112)$$

providing a simple energy estimate.

The propagation of the associated first-order constraints (4.106), namely

$$\mathcal{D}_{k\mu\nu} \equiv D_{k\mu\nu} - \partial_k\, g_{\mu\nu} = 0, \qquad (4.113)$$

will be given, allowing for (4.110), by

$$\partial_t\, \mathcal{D}_{k\mu\nu} - \beta^r\, (\partial_r\, \mathcal{D}_{k\mu\nu} - \partial_k\, \mathcal{D}_{r\mu\nu}) = \mathcal{B}_k{}^r\, D_{r\mu\nu} - \mathcal{B}_r{}^r\, D_{k\mu\nu}\,. \qquad (4.114)$$

As discussed before, the subsidiary system (4.114) is just weakly hyperbolic along any space direction $\boldsymbol{n}$ orthogonal to the shift vector (such that $\beta^n = 0$). Here again, constraint-damping terms may be added to (4.110), namely

$$\partial_t D_{k\mu\nu} + \partial_l [-\beta^l D_{k\mu\nu} + \delta^l_k \, \alpha \, Q_{\mu\nu}] = B_k{}^l \, D_{l\mu\nu} - trB \, D_{k\mu\nu} - \kappa\alpha \, \mathcal{D}_{k\mu\nu} \,, \quad (4.115)$$

so that the linearly growing constraint-violation modes arising from the principal part in (4.114) can be kept under control (we have set here $\eta = \kappa \, \alpha$).

Note, however, that the addition of the last term in (4.115) is affecting the principal part. The symmetric-hyperbolic character of the generalized harmonic system is lost in the process. This can be cured by adding a corresponding term in the evolution equation (4.109). The final equation can be written as (see [28] for more details):

$$\begin{aligned} \partial_t \, Q_{\mu\nu} &= \beta^k \, \partial_k Q_{\mu\nu} - \alpha \, \gamma^{ij} \partial_i D_{j\,\mu\nu} - \alpha \, (\partial_\mu H_\nu + \partial_\nu H_\mu - 2\Gamma_{\rho\mu\nu} \, H^\rho) \\ &+ 2 \, \alpha \, (\gamma^{ij} \, D_{i\mu\rho} D_{j\,\nu}{}^\rho - Q_{\mu\rho} Q_\nu{}^\rho - \Gamma_{\mu\rho\sigma} \Gamma_\nu{}^{\rho\sigma}) \\ &- \frac{\alpha}{2} \, n^\rho n^\sigma Q_{\rho\sigma} Q_{\mu\nu} + \alpha \, \gamma^{ij} \, D_{i\mu\nu} Q_{j\,\rho} n^\rho - 16\pi \, \alpha \, (T_{\mu\nu} - \frac{T}{2} \, g_{\mu\nu}) \\ &+ \kappa \, \beta^k (D_{k\mu\nu} - \partial_k \, g_{\mu\nu}) \end{aligned} \quad (4.116)$$

4.3.3 First-order Z4 formalism

As a second example, let us consider the first-order version of the Z4 system (3.80), (3.81), (3.82), and (3.83) in the general case, beyond the normal coordinates expressions (4.19), (4.20), (4.21), (4.22), (4.23), (4.24), and (4.25). For further convenience, we will express it in a balance law form, so that the principal part is flux conservative, namely

$$\partial_t \, \mathbf{u} + \partial_k \mathbf{F}^k(\mathbf{u}) = \mathbf{S}(\mathbf{u}) \,. \quad (4.117)$$

The evolution equations for the metric coefficients (4.86) and (4.88) and for the space derivatives quantities (4.95), (4.96), and (4.97) are in the required form (4.117). Concerning the remaining equations, we have

$$\begin{aligned} F^k[K_{ij}] &= -\beta^k \, K_{ij} + \alpha \, \lambda^k_{ij} & (4.118) \\ F^k \, [\Theta] &= -\beta^k \, \Theta + \alpha \, V^k & (4.119) \\ F^k[Z_i] &= -\beta^k \, Z_i + \alpha \, (\delta^k_i (trK - \Theta) - K^k_i) & (4.120) \\ & \qquad + \zeta' \, (B_i{}^k - \delta^k_i \, trB) \,, & \end{aligned}$$

where we keep using the shorthand (4.56), and

$$\lambda^k_{ij} \equiv D^k{}_{ij} + \frac{1}{2}\,\delta^k_i (A_j - D_j + 2V_j) + \frac{1}{2}\,\delta^k_j (A_i - D_i + 2V_i)$$
$$-\frac{1+\zeta}{2}\,(D_{ij}{}^k + D_{ji}{}^k - \delta^k_i E_j - \delta^k_j E_i)\,. \qquad (4.121)$$

Note that we have introduced in (4.120) an extra term, with a second ordering parameter $\zeta\,'$, which will contribute to the flux divergence in (4.117) only through the violations of the shift-related constraints $\mathcal{B}_k{}^i$.

The source terms $\mathbf{S}(\mathbf{u})$ do not belong to the principal part and will be displayed later. Let us focus for the moment on the hyperbolicity analysis, by selecting a specific space direction $\boldsymbol{n}$, so that the corresponding characteristic matrix is

$$\mathbf{A}^n = \frac{\partial\,\mathbf{F}^n}{\partial\,\mathbf{u}}\,. \qquad (4.122)$$

We get the following eigenfields, independently of the gauge choice, in (4.88):

- **Transverse derivatives**: $A_\perp$, $B_\perp{}^i$, $D_{\perp ij}$, propagating along the normal lines (characteristic speed $-\beta^n$).
- **Light-cone eigenfields**, given by the pairs

$$F^n[\,D_{n\perp\perp}\,] \pm F^n[\,K_{\perp\perp}\,] \qquad (4.123)$$
$$-F^n[\,Z_\perp\,] \pm F^n[\,K_{n\perp}\,] \qquad (4.124)$$
$$F^n[\,V_n\,] \pm F^n[\,\Theta\,], \qquad (4.125)$$

with characteristic speed $-\beta^n \pm \alpha$, respectively.

We cannot complete the analysis without specifying the lapse and shift prescriptions (4.88). In the spirit of the 3+1 formalism, the time slicing (lapse choice) is independent of the choice of space coordinates or, equivalently, the time lines (shift choice). Let us assume for the moment that we keep the singularity-avoidant lapse prescription (4.5), namely

$$Q = f\,(tr\,K - 2\,\Theta)\,, \qquad (4.126)$$

where we have chosen the second gauge parameter $m\ =\ 2$ here just for simplicity. We get in this way the **lapse-related eigenfields**

$$F^n[\,A_n\,] \pm \sqrt{f}\,F^n[\,Q\,], \qquad (4.127)$$

with characteristic speed $-\beta^n \pm \sqrt{f}\alpha$, respectively (gauge speed). The harmonic slicing case is recovered by taking $f = 1$.

There still remain some dynamical fields (trD , $D_{nn\perp}$, trB , $B_{n\perp}$) which depend on the shift prescription. Dynamical shift conditions have been considered in the literature, either by an algebraic prescription of Q^i [29–31], or by providing an independent evolution equation for this quantity, like the 'gamma-driver' condition [32, 33], which is currently used in the BSSN framework. Let us remark here that the shift prescription determines the behavior of the congruence of time lines, so that coordinate singularities (like shell crossing) can originate from inappropriate shift choices.

We will stay safe here, by considering the harmonic shift prescription, namely

$$Q_i = \alpha\,(A_i - D_i + 2V_i)\,. \tag{4.128}$$

The term 'harmonic shift' comes from the fact that this condition amounts to (the space part of) the prescription (3.76), which ensures the vanishing of the gauge sources components H_i . Condition (4.128) completes the dynamical system and allows to identify the **shift-related eigenfields**:

$$F^n[\,B_{ni}\,] \pm F^n[\,Q_i\,], \tag{4.129}$$

with characteristic speed $-\beta^n \pm \alpha$, respectively.

Allowing for these results, it follows that the full system is strongly hyperbolic for these specific gauge choices. Modifying the selected shift or lapse conditions, however, would require a new scrutiny of the characteristic cones generated by (4.129) or (4.127), respectively.

We give finally for completeness the source terms matching the fluxes displayed in (4.118), (4.119), and (4.119):

$$\begin{aligned} S(K_{ij}) &= -K_{ij}\; trB + K_{ik}\; B_j{}^k + K_{jk}\; B_i{}^k \\ &+ \alpha\,\{\, \frac{1}{2}\,(1+\xi)\,[\,-A_k\,\Gamma^k{}_{ij} + \frac{1}{2}(A_i\,D_j + A_j\,D_i)\,] \\ &+ \frac{1}{2}\,(1-\xi)\,[\,A_k\,D^k{}_{ij} - \frac{1}{2}\,\{A_j\,(2\,E_i - D_i) + A_i\,(2\,E_j - D_j)\} \\ &+ 2\,(D_{ir}{}^m\,D^r{}_{mj} + D_{jr}{}^m\,D^r{}_{mi}) - 2\,E_k\,(D_{ij}{}^k + D_{ji}{}^k)\,] \\ &+ (D_k + A_k - 2\,Z_k)\,\Gamma^k{}_{ij} - \Gamma^k{}_{mj}\,\Gamma^m{}_{ki} - (A_i\,Z_j + A_j\,Z_i) \\ &- 2\,K^k{}_i\,K_{kj} + (trK - 2\,\Theta)\,K_{ij}\,\} - 8\pi\,\alpha\,[\,S_{ij} - \frac{1}{2}\,(trS - \tau)\,\gamma_{ij}\,] \end{aligned} \tag{4.130}$$

$$\begin{aligned} S(Z_i) &= -Z_i\,trB + Z_k\,B_i{}^k - 8\pi\,\alpha\,S_i \\ &+ \alpha\,[\,A_i(trK - 2\,\Theta) - A_k K^k{}_i - K^k{}_r\Gamma^r{}_{ki} + K^k{}_i(D_k - 2Z_k)\,] \end{aligned} \tag{4.131}$$

$$\begin{aligned} S(\Theta) &= -\Theta\;trB + \frac{\alpha}{2}\,[2A_k\,(D^k - E^k - 2\,Z^k) + D_k{}^{rs}\,\Gamma^k{}_{rs} - D^k(D_k - 2\,Z_k) \\ &- K^k{}_r\,K^r{}_k + trK\,(trK - 2\,\Theta)] - 8\pi\,\alpha\,\tau\,. \end{aligned} \tag{4.132}$$

4.4 Boundary conditions

Most numerical relativity simulations are devised to approximate the time evolution of the dynamical fields starting from data given on an initial time slice: the general relativistic Cauchy or initial-value problem (IVP). The theoretical formalisms we have described so far are built with the objective of getting a well-posed Cauchy problem. This ensures, at the continuum level, that the solution is unique and depends smoothly on the initial data. A well-posed Cauchy problem is also a necessary condition for the existence of stable numerical algorithms that transpose the same property at the discrete level: the time evolution of the selected initial data must provide a sound approximation to the corresponding solution (the accuracy must improve with increasing numerical resolution).

The numerical applications we have presented up to now were dealing just with periodic solutions. This means that we could identify the first points of the numerical grid with the last ones along every axis: from the continuum point of view, our computational arena has the topology of a three-torus. In this way, there is no genuine 'first' or 'last' point: we can always rotate the torus along one symmetry axis, so that the chosen point gets surrounded by the number of neighbors required for a given computational stencil. As a consequence, only initial data are required to produce a solution, both at the continuum and the discrete levels.

In physical applications, however, we will rarely find this three-torus topology. Three-dimensional finite difference grids usually start at some point and end at some other one along every axis, and this fact poses a serious problem both at the continuum and at the discrete level:

- From the discrete point of view, the numerical stencil that is being used for the 'interior' points needs to be modified at end points, or even at next-to-end points, where the required number of neighbors is not available inside the numerical grid.
- From the continuum point of view, the computational domain has a border, so that information can cross it in both senses. Of course, incoming information cannot be obtained from inside, so it must be specified through additional conditions. These 'boundary conditions' are relevant to the well-posedness of the system: the pure initial-value problem is then transformed into a initial boundary problem (IBVP).

4.4.1 Algebraic boundary conditions

In order to clarify these points, let us focus for the moment in the one-dimensional (1D) case. We see in Fig. 4.1 the domain of dependence of a boundary point P, displayed as a shaded cone. It is clear that the regions outside the numerical grid do have causal influence on P, so that extra physical information is required at the boundary. From the numerical point of view, the stencil at P must be one-sided (like the one displayed as a dotted triangle) instead of the centered stencil discussed before: this means that the discretization algorithm must be changed at P.

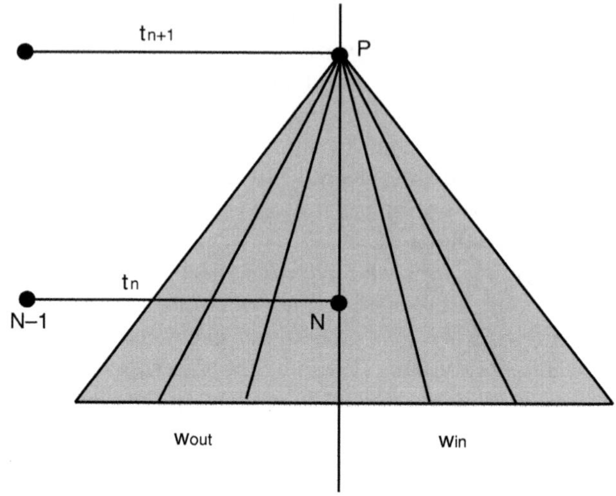

Fig. 4.1 The domain of dependency of a given boundary point P is shown as a *gray cone*. The outgoing components w^{out} are contained into the *left half-cone*, so that their values can be computed consistently by using a one-sided (upwind) stencil. The ingoing components w^{in} belong instead to the *right half-cone*, which is outside the numerical grid: they must be provided by the boundary conditions.

To be more specific, let us consider a 1D strongly-hyperbolic flux-conservative system (we neglect source terms for the moment)

$$\partial_t \, \mathbf{u} + \partial_x \, \mathbf{F} = 0 \,. \tag{4.133}$$

The characteristic matrix

$$\mathbf{A} = \frac{\partial \mathbf{F}}{\partial \mathbf{u}} \tag{4.134}$$

can be fully diagonalized and the set of dynamical fields $\mathbf{u}$ can then be decomposed in the basis of eigenfields $\mathbf{w}$ given by (4.66).

We will classify the characteristic eigenfields into two subsets:

- **Incoming** eigenfields $\{w_r^{\mathrm{in}}\}$: the ones with characteristic speed pointing inside the computational region ($\lambda_r < 0$).
- **Outgoing** eigenfields $\{w_r^{\mathrm{out}}\}$: the ones with characteristic speed pointing outside the computational region, or tangent to the boundary ($\lambda_r \geq 0$).

To justify the use of the 'incoming' and 'outgoing' terms, let us consider again the simplest case, when the characteristic matrix (4.134) has constant coefficients. Then, the original system (4.133) can be decoupled into a set of independent advection equations (4.68) for the characteristic fields, so that the propagation behavior is like the one shown in Fig. 4.1. Simple causality considerations lead then to the two main requirements for physically sound boundary conditions:

- Boundary conditions must provide the values of all the incoming eigenfields. Otherwise we will lose the uniqueness of the solution (not enough information is given) and this will result in numerical instabilities.
- Boundary conditions must not restrict the values of the outgoing eigenfields. This is because the domain of dependence of these fields is fully contained into the computational region, so that their values are completely determined by inside information. Any further condition would be either redundant or inconsistent.

The prototype of a boundary condition verifying these two main requirements is provided by the simple algebraic relationship

$$w_r{}^{\text{in}} = \sum_s M_{rs}\, w_s{}^{\text{out}} + C_r\,, \tag{4.135}$$

where the coefficients M_{rs} and C_r are independent of the dynamical fields. We will explore algebraic conditions (4.135) in more detail in what follows.

4.4.2 Energy methods

We have shown in Sect. 4.2.2 how to obtain an energy estimate $\overline{E}$ for any 1D strongly hyperbolic system. The time evolution of such estimate is given by (4.73). A closer look suggests the following decomposition in terms of incoming and outgoing eigenfields:

$$\partial_t\, \overline{E} = -\left[\sum_r \lambda_r^{\text{in}} (w_r^{\text{in}})^2 + \sum_s \lambda_s^{\text{out}} (w_s^{\text{out}})^2\right]_a^b , \tag{4.136}$$

where a and b are the limits of the computational domain.

It is clear that the outgoing terms contribution goes in the right sense $(\,\partial_t\, \overline{E} \leq 0\,)$, whereas incoming terms contribute instead in the wrong one $(\,\partial_t\, \overline{E} > 0\,)$ because their characteristic speed is positive at the first point a and negative at the last point b.

Now we are in a position to analyze the effect of different choices in the generic boundary condition (4.135):

- The most conservative one would be to suppress all the incoming fields at the boundaries, that is,
$$w_s^{\text{in}} = 0 \qquad \forall s \tag{4.137}$$
(maximally dissipative boundary condition). In this case it is clear that the energy estimate $\overline{E}$ will diminish with time at the maximum rate.
- The next safe choice would be to relate every incoming field at the boundary with an outgoing one, grouped by pairs, that is,

$$w_s^{\rm in} = \sigma \; w_r^{\rm out} \,, \tag{4.138}$$

so that the corresponding terms in the sum (4.136) can be grouped as

$$(\lambda_r^{\rm out} + \sigma^2 \lambda_s^{\rm in}) \; (w_r^{\rm out})^2 \,, \tag{4.139}$$

which will have the right sign provided that the 'reflection coefficient' σ is small enough, that is,

$$\lambda_r^{\rm out} + \sigma^2 \lambda_s^{\rm in} \geq 0 \,. \tag{4.140}$$

- An interesting particular case of the former one is when the corresponding outgoing speed $\lambda_r^{\rm out}$ is zero. Then, there is no 'small enough' non-trivial choice of σ that could verify (4.140). One has instead the contribution

$$\sigma^2 \lambda_s^{\rm in} \; (w_r^{\rm out})^2 \,, \tag{4.141}$$

  which goes always with the wrong sign. Notice, however, that this term is constant in time ($\lambda_r^{\rm out}$ is actually zero), so that it will produce a linear increase of the energy estimate $\overline{E}$ in the worst case.
- As an alternative, one could just specify the incoming fields at the boundaries in terms of some given functions, namely

$$w_s^{\rm in} = C_s(t) \,, \tag{4.142}$$

  so that a convenient upper bound for the growth of the energy estimate can be obtained by a suitable choice of the functions $C_s(t)$ at the boundaries.

Any of the above conditions is strong enough in the 1D case, so that this 'energy method' can be used for proving that the IBVP is well-posed, as was the corresponding IVP, for any strongly hyperbolic system.

In the multidimensional case, however, the corresponding energy estimates (4.78) can be obtained just for symmetric-hyperbolic systems: strong hyperbolicity is not enough. We have seen in Sect. 4.2.3 how to get an energy estimate in the symmetric-hyperbolic case, which time variation is given by (4.79) in terms of a surface integral over the boundary of the computational domain. As discussed in Sect. 4.2.2, the term appearing in each surface integral is a scalar with respect to the orthogonal transformations (4.77). This scalar can be computed in the basis of characteristic fields associated with the normal $\boldsymbol{n}$ to the boundary surface. Note that the outward unit normal $\boldsymbol{n}$ at the boundary must be defined at corner points (see Fig. 4.2).

The corresponding characteristic matrix $\mathbf{A}^n$ will take the diagonal form:

$$\mathbf{A}^n = {\rm diag}(\{\lambda_r^{\rm in}\}, \; \{\lambda_s^{\rm out}\}) \,, \tag{4.143}$$

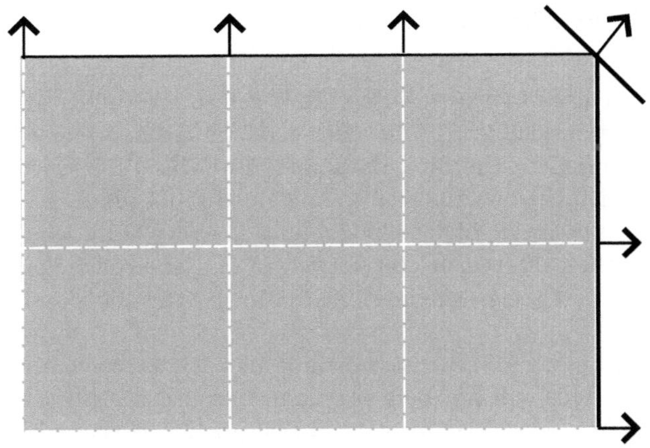

Fig. 4.2 The outward normal to the boundaries of a 2D grid are shown. A normal direction must be chosen at *corner points*. Note that the decomposition into incoming and outgoing eigenfields is direction dependent, so that there are physical 'in' and 'out' modes along every direction, not just normal to the boundary surface.

so that one finally gets the same kind of result than in the 1D case

$$\partial_t\, \overline{E} = -\oint \left[\, \sum_r \lambda_r^{\rm in} (w_r^{\rm in})^2 + \sum_s \lambda_s^{\rm out} (w_s^{\rm out})^2 \right] \, dS\,. \tag{4.144}$$

This means that the energy method can be applied to devise boundary conditions for multidimensional symmetric-hyperbolic systems in the same way as in the 1D case, as discussed before. Some words of caution are convenient at this point:

- Energy estimates can provide sufficient conditions for a well-posed IBVP in the symmetric-hyperbolic case. But, if the energy metric is not unique, the resulting conditions may not be necessary.
- The arguments presented here apply just to the continuum level. Sufficient conditions at the discrete level will follow only if the numerical algorithm is able to preserve the integration-by-parts rule, which plays a key role in these developments. This restricts the choice of discrete algorithms, both at the interior points and at the boundary ones (see for instance [34]).
- The results presented here hold only for pure flux-conservative systems (no source terms) and for the linearized case, where the coefficients of the characteristic matrices are constant. None of these requirements is fulfilled in numerical relativity applications: one must then use energy estimates just as a guideline. The stability of the resulting boundary conditions can be rather checked numerically, as we will see below.

4.4.3 Robust stability test

Let us go back here to the robust stability test. We will use again a small (20^3 nodes) numerical grid. The centered algorithm we used in Sect. 3.2 will still be applied to the interior points. But this time, with non-periodic boundary conditions, we will use also one-sided finite difference algorithms to compute (a first estimate of) the field values at every boundary surface. To be more specific, the centered discretization D^0 for the space derivative will be replaced at the first (resp. the last) grid point by the one-sided discretization D^- (resp. D^+).

As it is well known [35], these 'upwind' algorithms are unstable for the incoming modes. This will not be a problem, because we will use the algebraic boundary conditions (4.138) for replacing a posteriori the values of precisely those incoming modes. Only the outgoing modes, for which the upwind algorithms are stable, will contribute then to the final boundary values.

We will choose for this test the (strongly hyperbolic) Z4 system in normal coordinates. This is in order to get advantage of the zero shift case, where the characteristic lines corresponding to transverse space derivatives are tangent to the selected boundary. In all cases we will take $f = 1$, $m = 2$ for the gauge parameters. In this way, we can cover the symmetric-hyperbolic case $\zeta = -1$ and the plain strongly hyperbolic cases for any other values of the ordering parameter.

Starting with the symmetric-hyperbolic case, we see in Fig. 4.3 the time evolution of $\|trK\|_\infty$, that is the maximum of the absolute value of trK,

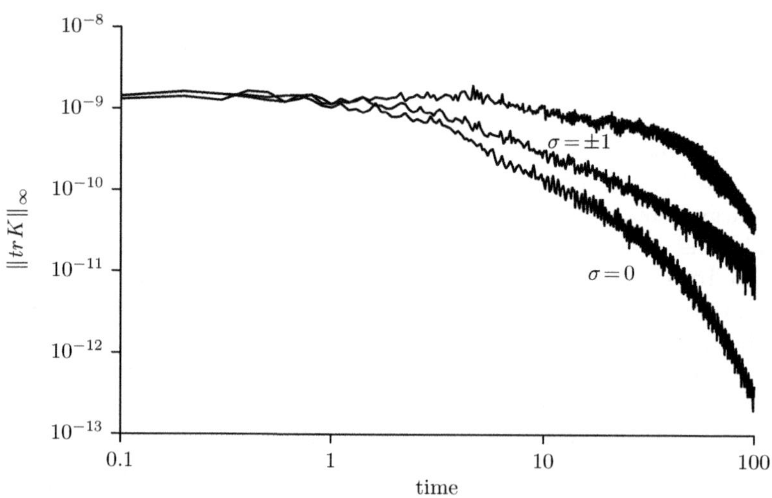

Fig. 4.3 Time evolution of the maximum of the absolute value of trK, which is one of the main contributions to the energy estimate. For the maximally dissipative case $\sigma = 0$, the expected decreasing is clearly seen. The unexpected decreasing of the $\sigma = \pm1$ case is due to numerical dissipation, caused by the use of the first-order upwind algorithm at the boundary points.

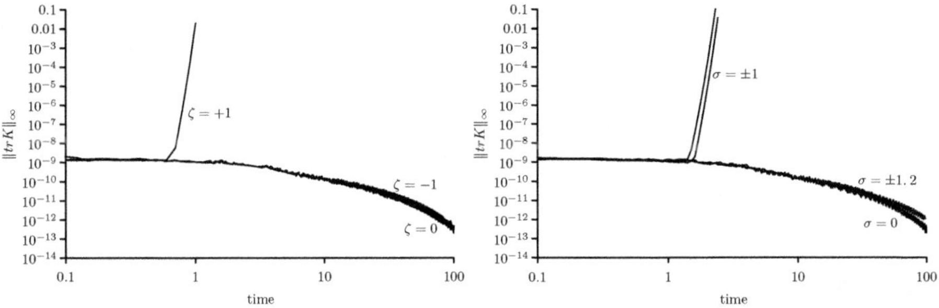

Fig. 4.4 Same as Fig. 4.3 for different values of ζ and σ. The plots in the *left panel* correspond to different choices of the ordering parameter in the maximally dissipative case ($\sigma = 0$). It follows that the $\zeta = +1$ choice of the ordering parameter is unstable. Conversely, the plots in the *right panel* correspond to different values of the reflection coefficients in the symmetric ordering case ($\zeta = 0$). The extreme choices $\sigma = \pm 1$ are unstable, while other choices, closer to the maximally dissipative one, still work.

which is one of the main contributions to the energy estimate (4.81). For the maximally dissipative choice $\sigma = 0$, the expected decreasing is clearly seen. The case $\sigma = \pm 1$ deserves a further comment, because every non-zero eigenvalue appears twice, with different signs (zero shift case), that is,

$$\lambda^{\text{in}} = -\lambda^{\text{out}}, \tag{4.145}$$

so that the resulting terms (4.139) do not contribute to the energy decreasing when $\sigma^2 = 1$. Then, a constant value of the energy estimate would be expected at the continuum level. There are two alternative ways of explaining the observed decreasing when $\sigma = \pm 1$ in Fig. 4.3:

- As the effect of the non-linear source terms and/or the non-constant coefficients in the characteristic matrices involved. This could also explain the different behavior of the $\sigma = +1$ and the $\sigma = -1$ cases. This is very unlikely, because the noise level we are using here is so low that non-linear contributions can be safely neglected.
- As the effect of numerical dissipation. This does not come from the numerical algorithm used at interior points, which does not show so much dissipation (see for instance Fig. 3.3). This must be due instead to the use of the first-order upwind method, which is known to be highly dissipative, at the boundary points.

We can check now the plain strongly-hyperbolic case. As explained before, the only change will be to consider different values of the ordering parameter ζ, so that the evolution system is just strongly hyperbolic, but not symmetric hyperbolic. We show again in Fig. 4.4 the time evolution of the maximum of the absolute value of trK.

In the maximally dissipative case (left panel), we see that the choice $\zeta = +1$ of the ordering parameter is unstable, whereas the two other choices $\zeta = 0, -1$ are stable and behave in the same way. This provides a counterexample for the common belief that maximally dissipative boundary conditions 'should always work.' In the (strongly hyperbolic) $\zeta = 0$ case, shown in the right panel, we see that the 'extreme' values $\sigma = \pm 1$ of the reflection coefficients lead to code crashing, whereas other choices, closer to the maximally dissipative one $\sigma = 0$, work fine, showing the same decreasing behavior as in the symmetric-hyperbolic case.

References

1. T. De Donder, *La Gravifique Einstenienne,* Gauthier-Villars, Paris (1921).
2. T. De Donder, *The Mathematical Theory of Relativity*, Massachusetts Institute of Technology, Cambridge (1927).
3. K. Lanczos, Ann. Phys. **13**, 621 (1922).
4. K. Lanczos, Z. Phys. **23**, 537 (1923).
5. V. A. Fock, *The Theory of Space, Time and Gravitation*, Pergamon, London (1959).
6. L. E. Kidder, M. A. Scheel and S. A. Teukolsky, Phys. Rev. **D64**, 064017 (2001).
7. O. Sarbach and M. Tiglio, Phys. Rev. **D66**, 064023 (2002).
8. H. O. Kreiss and J. Lorentz, *Initial-Boundary Problems and the Navier-Stokes Equations*, Academic Press, New York (1989).
9. Y. Choquet-Bruhat and T. Ruggeri, Commun. Math. Phys. **89**, 269 (1983).
10. R. Arnowit, S. Deser and C. W. Misner. *Gravitation: An Introduction to Current Research*, ed. by L. Witten, Wiley, New York (1962). gr-qc/0405109.
11. C. Bona and J. Massó, Phys. Rev. Lett. **68**, 1097 (1992).
12. C. Bona, J. Massó, E. Seidel and J. Stela, Phys. Rev. Lett. **75,** 600 (1995).
13. K. Alvi, Class. Quantum Grav, **19**, 5153 (2002).
14. M. Holst, et al., Phys. Rev. **D70**, 084017 (2004).
15. O. Brodbeck, S. Frittelli, P. Huebner and O. A. Reula, J. Math. Phys. **40**, 909 (1999).
16. S. Frittelli and O. A. Reula, Commun. Math. Phys. **166**, 221 (1994).
17. S. Frittelli and O. A. Reula, Phys. Rev. Lett. **76**, 4667 (1996).
18. A. Anderson and J. W. York, Jr., Phys. Rev. Lett. **82**, 4384 (1999).
19. S. D. Hern, Numerical Relativity and Inhomogeneous Cosmologies. Ph. D. Thesis, gr-qc/0004036 (2000).
20. A. Abrahams, A. Anderson, Y. Choquet-Bruhat and J. W. York, Phys. Rev. Lett. **75**, 3377 (1995).
21. H. Friedrich, Class. Quantum. Grav. **13**, 1451 (1996).
22. M. H. P. van Putten and D. M. Eardley, Phys. Rev. **D53**, 3056 (1996).
23. M. H. P. van Putten, Phys. Rev. **D55**, 4705 (1997).
24. G. Nagy, O. Ortiz and O. Reula, Phys. Rev. **D70**, 044012 (2004).
25. C. Bona, T. Ledvinka, C. Palenzuela and M. Žáček, Phys. Rev. **D69**, 064036 (2004).
26. C. Bona, T. Ledvinka, C. Palenzuela and M. Žáček, Phys. Rev. **D67**, 104005 (2003).
27. F. Pretorius, Phys. Rev. Lett. **95**, 121101 (2005).
28. L. Lindblom et al., Class. Quantum Grav, **23**, S447 (2006).
29. L. Lindblom and M. A. Scheel, Phys. Rev. **D67**, 124005 (2003).
30. C. Bona and C. Palenzuela, Phys. Rev. **D69**, 104003 (2004).
31. D. Alic, C. Bona and C. Bona-Casas, Phys. Rev. **D** (2009). ArXiv:0811.1691
32. M. Alcubierre and B. Brügmann, Phys. Rev. **D63**, 104006 (2001).
33. M. Alcubierre et al., Phys. Rev. **D67**, 084023 (2003).
34. P. Olsson, Math. Comput. **64**, 1035 (1995).
35. W. H. Press, B. P. Flannery, S. A. Teukolsky and W. T. Veterling, *Numerical Recipes*, Cambridge University Press, Cambridge (1989).

Chapter 5
Numerical Methods

A numerical relativity code consists in two main ingredients: the evolution system and the discretization algorithm. Up to now, we have focused in the evolution formalism. The strong hyperbolicity requirement is a requisite for a well-posed system *at the continuum level.* Also, the subsidiary system, governing constraint deviations, has been studied at the continuum level, where we have seen how the subset of true Einstein's solutions can become an attractor for extended (constraint-violating) solutions.

But all these interesting properties are not guaranteed at the discrete level, unless we implement a suitable numerical algorithm. There is no unique prescription providing an optimal choice in all cases. If we consider just spacetime evolution, the smoothness of the metric coefficients allows using either finite difference or spectral methods. Finite differencing is systematically used in numerical relativity for time discretization and it is also the mainstream option for space discretization. We will discuss this approach in the first section.

The case of matter evolution is different. Fluid dynamics is a genuine non-linear domain, where one can get shock propagation arising even from smooth initial data. In this framework, the mainstream option for the space discretization is using finite volume methods, in many variants, as we will discuss in the following sections, including many standard numerical tests from computational fluid dynamics (CFD).

Moreover, both in the finite difference and the finite volume approaches, we are forced to deal at the discrete level with a finite numerical mesh. This usually requires implementing some boundary conditions, which must be consistent at the continuum level and stable at the numerical one. These issues will be discussed at the end of the chapter, including some numerical stability tests.

Bona, C. et al.: *Numerical Methods*. Lect. Notes Phys. **783**, 109–142 (2009)
DOI 10.1007/978-3-642-01164-1_5

5.1 Finite difference methods

An elementary introduction to finite differencing has been already included in Sect. 3.2.1, in the context of the robust stability test. We will provide here a more complete treatment, by following the method-of-lines (MoL) approach, which allows to separate the space and time-discretization processes. The current practice is using finite differences for the time discretization, even in cases in which the space discretization is dealt with a finite volume approach.

5.1.1 Accuracy and stability

Let us start by considering the one-dimensional (1D) advection equation

$$\partial_t\, u + \lambda\, \partial_x\, u = 0\,. \tag{5.1}$$

This can provide a good 1D prototype, as we have seen in the previous chapter how the dynamics of (the principal part of) any strongly hyperbolic system can be understood as the superposition of one advection equation for every characteristic field.

The simplest finite difference algorithm for (5.1) is given by

$$D_t^+\, u + \lambda\, D_x^0\, u = 0\,, \tag{5.2}$$

where we keep using the notations introduced in Sect. 3.2.1 for the differences operators D. In this case

$$D_t^+\, u = (u_i^{(n+1)} - u_i^{(n)})/\Delta t \tag{5.3}$$

$$D_x^0\, u = (u_{i+1}^{(n)} - u_{i-1}^{(n)})/(2\Delta x)\,, \tag{5.4}$$

that is, the Euler (forward) time step in time combined with the centered space differencing (FTCS algorithm):

$$u_i^{(n+1)} = u_i^{(n)} - \lambda\, \Delta t\, D_x^0\, u^{(n)}\,. \tag{5.5}$$

We can check the accuracy of the discretization procedure by a standard Taylor development around u_i^n, namely

$$D_t^+\, u \sim \partial_t\, u + \frac{1}{2}\, \Delta t\; \partial_{tt}^2\, u + \cdots \tag{5.6}$$

$$D_x^0\, u \sim \partial_x\, u + \frac{1}{6}\, (\Delta x)^2\; \partial_{xxx}^3\, u + \cdots. \tag{5.7}$$

It follows that the FTCS algorithm (5.5) is second-order accurate in space (third-derivative leading error), but just first-order accurate (second-derivative leading error) in time. Note that these orders refer to the accuracy of the solutions: the accuracy in the first derivatives (5.6) and (5.7) is actually reduced by one order. Note also that we can use the advection equation (5.1) in order to relate space and time derivatives. It is clear then that the leading error is given by

$$\frac{1}{2}\,\Delta t\;\partial^2_{tt}\,u \qquad \Leftrightarrow \qquad \frac{\lambda^2}{2}\,\Delta t\;\partial^2_{xx}\,u. \tag{5.8}$$

We can interpret then that the discrete scheme (5.2) is approximating rather the modified equation

$$\partial_t\;u + \lambda\,\partial_x\;u = -\frac{\lambda^2}{2}\,\Delta t\;\partial^2_{xx}\,u \tag{5.9}$$

(modified equation approach). An elementary Fourier analysis, considering a single mode of the form

$$u = \xi(t)\,\mathrm{Sin}(x - \lambda\,t) \tag{5.10}$$

shows that the extra term in (5.9) is of a dissipative type, but with the wrong sign. It describes an explosive behavior, with an exponential growth of the mode amplitude $\xi(t)$. It follows that the simple Euler step discretization (5.3) is unstable, no matter how small we take the time step.

This elementary Fourier analysis actually shows that even-derivatives leading error terms amount to numerical dissipation, which can either smooth out or blow up the numerical solution, depending on the sign in the modified equation. Odd-derivatives leading error terms, on the contrary, amount to numerical dispersion, affecting the propagation speed. We can see this by assuming for a moment that the space discretization leading error (5.7) is the dominant one. The modified equation would read then

$$\partial_t\,u + \lambda\,\partial_x\,u = -\frac{1}{6}\,\lambda\,(\Delta x)^2\;\partial^3_{xxx}\,u\,. \tag{5.11}$$

For a single Fourier mode (5.10), this can be reduced to the modified advection equation

$$\partial_t\,u + \lambda\,(1 - \frac{1}{6}\,\Delta x^2)\;\partial_x\,u = 0\,, \tag{5.12}$$

which amounts to a second-order error in the propagation speed.

Coming back to the time derivative discretization, it is clear that a stable method can be obtained by taking rather

$$D_t^-\;u + \lambda\,D_x^0\,u = 0\,, \tag{5.13}$$

so that the numerical dissipation gets the right sign in the modified equation. This would lead to the BTCS (bakward-in-time, centered-in-space) algorithm, namely

$$u_i^{(n+1)} + \lambda\, \Delta t\, D_x^0\, u^{(n+1)} = u_i^{(n)} \,. \tag{5.14}$$

This is an implicit algorithm, in the sense that the individual values at the next time level can only be obtained after some matrix-inversion process involving the whole computational mesh. This is numerically too expensive in multidimensional applications, so that explicit methods are currently preferred.

Stable explicit methods, like the Lax–Wendroff [1] or the Mac–Cormack ones [2] can be obtained by modifying the space discretization in order to get extra dissipative terms that cancel out the unstable contribution for the forward time discretization (5.3). The resulting second-order-accurate schemes can be found in standard books [3]. But the need to balance the time and space discretization is a complication when going beyond second-order accuracy. This is why we prefer to adopt a more powerful strategy, as described in the next section.

5.1.2 The method of lines

The method of lines (MoL) [4] is the generic name of a family of discretization methods in which time and space variables are dealt with separately. This is in keeping with the 3+1 framework, where the natural way of time discretization is by finite differences whereas one would like to keep all the options open for space discretization: finite differences, finite volume, or even spectral methods.

To illustrate the idea, let us consider a 'semi-discrete' system in which only the time coordinate is discretized, whereas space derivatives are kept at the continuum level. The evolution of the array $\mathbf{u}$ of dynamical fields is written as

$$\partial_t\, \mathbf{u} = \mathbf{S} \,, \tag{5.15}$$

where the right-hand-side array $\mathbf{S}$ contains the remaining terms in the evolution equations, including the space derivative ones. In this way, we are disguising in (5.15) the original system of partial differential equations (PDE) as a system of ordinary differential equations (ODE), assuming that we will manage to compute the right-hand-side term $\mathbf{S}$ at every time level, but ignoring for the moment the details.

This 'black box' approach allows us to apply the well-known ODE discretization techniques to get the required time resolution, using the Euler step (forward time difference)

$$\mathbf{u}^{(n+1)} = \mathbf{u}^{(n)} + \Delta t\; \mathbf{S}(t_n, \mathbf{u}^{(n)}) \tag{5.16}$$

as the basic building block for advanced multi-step methods, like the modified-midpoint or Runge–Kutta algorithms [3].

A simple Runge–Kutta algorithm would read for instance

$$\mathbf{S}^{(n)} = \mathbf{S}(t_n, \mathbf{u}^{(n)}) \tag{5.17}$$

$$\mathbf{u}^{*} = \mathbf{u}^{(n)} + \Delta t\ \mathbf{S}^{(n)} \tag{5.18}$$

$$\mathbf{u}^{(n+1)} = \frac{1}{2}\ \mathbf{u}^{(n)} + \frac{1}{2}\ \mathbf{u}^{*} + \frac{\Delta t}{2}\ \mathbf{S}^{*}\,, \tag{5.19}$$

which is second-order accurate. We have seen in the past section how those second-order methods provide just first-order accurate propagation speeds (second-order error in λ). This is why we recommend using at least third-order accurate algorithms for time evolution. In the simulations presented in Sect. 3.2, the following third-order Runge–Kutta algorithm [5] was actually used:

$$\mathbf{u}^{*} = \mathbf{u}^{(n)} + \Delta t\, \mathbf{S}^{(n)} \tag{5.20}$$

$$\mathbf{u}^{**} = \frac{3}{4}\ \mathbf{u}^{(n)} + \frac{1}{4}\ \mathbf{u}^{*} + \frac{\Delta t}{4}\ \mathbf{S}^{*} \tag{5.21}$$

$$\mathbf{u}^{(n+1)} = \frac{1}{3}\ \mathbf{u}^{(n)} + \frac{2}{3}\ \mathbf{u}^{**} + \frac{2}{3}\ \Delta t\ \mathbf{S}^{**}\,. \tag{5.22}$$

Note that increasing the accuracy order comes at the price of increasing the number of intermediate steps. Also, for a given accuracy order, there are different Runge–Kutta algorithms, which are devised either for minimizing the number of operations or for maximizing the time step Δt allowed by the stability condition (corresponding to the Courant condition discussed in Sect. 3.2). The specific versions (5.19) and (5.22) have been rather selected for the sake of robustness. Note that the positivity of all coefficients means that the final result is a convex combination of successive Euler steps. These are called ‘strong stability preserving’ (SSP) algorithms, because the (strong) stability of the Euler step would be inherited by the complete multi-step method. Of course, the stability of the Euler step will depend on the selected space discretization, as we will discuss in what follows.

5.1.3 Artificial dissipation

Let us now focus on the space discretization part, and more specifically in the finite differences method. We will start by the simple semi-discrete algorithm for the advection equation (5.1):

$$\partial_t\, u + \lambda\, D^0_x\, u = 0\,. \tag{5.23}$$

Note that here the approach is just the opposite one. The black box is the time derivative term, which we will consider at the continuum (exact) level. Only the space part is now discretized.

We know from our previous analysis that the centered space discretization in (5.23) is second-order accurate. Moreover, the left–right symmetry of centered algorithms cancels out all even-order derivatives in the Taylor development. It means that the semi-discrete algorithm (5.23) is free of any numerical dissipation. This can be acceptable in simple problems, because (as we actually did in the simulations in Sect. 3.2) the dissipation coming from the (stable) time discretization (5.22) will ensure the stability of the combined, fully discrete, numerical schemes.

More robust algorithms are, however, required when dealing with steep-gradient profiles or when some spurious numerical modes need to be kept under control. The obvious alternative is to add an 'artificial dissipation' term to (5.23), inspired by our previous 'modified equation' analysis, which will provide the required stability, independent of the time discretization. The algorithm (5.23) uses a three-point stencil, so adding a second derivative term is the obvious choice, that is,

$$\partial_t\, u + \lambda\, D^0_x\, u - \eta\, \Delta x\, D^+_x\, D^-_x\, u = 0\,, \tag{5.24}$$

where the dissipation coefficient η must be positive. Note that, as the magnitude of η is not determined, the Δx power in the dissipation term is only relevant if one keeps the same dissipation coefficient, independent of the resolution (see also [6]).

The improved robustness of the semi-discrete algorithm (5.24) comes, however, with an unacceptable accuracy loss, when we get a second-order (dissipative) error. Robust high-accuracy algorithms can be nevertheless constructed by following the same pattern:

- Start with the $2s$-accurate centered difference operator C^{2s} for the first derivative (stencil size s, that is, $2s+1$ points). This can be obtained in a standard way, by combining centered first derivatives of different space intervals, namely

$$C^{2s} = 2 \sum_{k=1}^{s} \frac{(-1)^{k-1}(\,s\,!\,)^2}{(\,s+k\,)!\,(\,s-k\,)!}\; D^0_{(kx)}\,. \tag{5.25}$$

- Add a dissipative term of order $2s$, so that the resulting algorithm

$$\partial_t\, u_j = -\lambda\; C^{2s} u_j + (-1)^{s-1} \eta\, (\Delta x)^{2s-1} (D^+_x\, D^-_x\,)^s\, u_j \tag{5.26}$$

  is $(2s-1)$-order accurate in the solutions ($2s-2$ in the first derivatives).

The case $s = 1$ ($C^2 = D^0$) corresponds to the first-order accurate algorithm (5.24). In numerical relativity applications, a good cost-efficiency ratio is rather provided by the third-order accurate case $s = 2$ (five-points stencil), as we will see later in this chapter.

The artificial dissipation algorithms (5.26) can be easily generalized to systems. In the flux-conservative case

$$\partial_t \, \mathbf{u} + \partial_x \, \mathbf{F} = 0 \,, \tag{5.27}$$

we will have for every component

$$\partial_t \, u_j = -C^{2s} F_j + (-1)^{s-1} \eta \, (\Delta x)^{2s-1} (D_x^+ D_x^-)^s \, u_j \,. \tag{5.28}$$

We prefer rather to use a variant of this, which will be justified later in the finite volume context, namely

$$\partial_t \, u_j = -C^{2s} F_j + (-1)^{s-1} \, b \, (\Delta x)^{2s-1} (D_x^+ D_x^-)^{s-1} \, D_x^+ [\, \lambda_M \, D_x^- u_j \,] \,, \tag{5.29}$$

where we have noted here

$$\lambda_M = \max \, (\lambda_j , \, \lambda_{j-1}) \,. \tag{5.30}$$

and λ_j is the local spectral radius of the system (the maximum characteristic speed).

All these artificial dissipation algorithms can be generalized to the multidimensional case in a straightforward way, by considering one space direction at a time, so that no cross derivatives appear in the dissipation terms. The spectral radius appearing in (5.29) must correspond to the characteristic matrix along every selected direction. The algorithm (5.29) has shown its robustness even in demanding CFD tests, with interacting shocks in multidimensional cases [7]. We will present just some simple cases in the finite volume section in this chapter.

5.1.4 The gauge waves test-bed

Now we will test the finite differences algorithms presented in this chapter in a very peculiar situation: Minkowski spacetime endowed with a non-conventional gauge condition. To understand what we mean by this, let us start from Minkowski metric in the standard inertial coordinate system

$$ds^2 = -dt^2 + dx^2 + dy^2 + dz^2 \,, \tag{5.31}$$

and perform a general conformal transformation in the 2D sector spanned by the t, x coordinates, namely

$$ds^2 = H^2(t, x) \; (-dt^2 + dx^2) + dy^2 + dz^2 \,, \tag{5.32}$$

where $H(t,\ x)$ is an arbitrary function and the $(t,\ x)$ labels refer now to the transformed coordinates.

We can simulate here propagation along the x axis (in the positive sense, for instance), by taking

$$H(t,x) = h(x-t) \tag{5.33}$$

(notice that light speed along the x axis is still $v = 1$ in the transformed coordinates). The exact time evolution can then be easily obtained from any given initial profile

$$H(0,x) = h(x)\,. \tag{5.34}$$

Note that (5.33) implies

$$\alpha = h\,, \qquad K_{xx} = h'\,, \qquad A_x = h'/h\,, \qquad D_{xxx} = hh'\,. \tag{5.35}$$

The term 'gauge waves' is justified by the propagation behavior (5.33) and the fact that the only non-trivial eigenfields belong to the gauge sector (4.64). Note that the metric form (5.32) corresponds to a harmonic slicing condition ($f = 1$), so that we must also have $m = 2$ to ensure strong hyperbolicity. The ordering parameter ζ is irrelevant here because only the x derivatives provide non-trivial contributions. Note also that in this particular case the normal space coordinates are also harmonic coordinates. This is why that particular line element has been included in a battery of cross-comparison test-beds, suitable for most of the current evolution formalisms ('apples with apples' campaign, see [8, 9]).

We will consider for this test the 'gauge waves' line element with the following profile:

$$H = 1 - A\ \sin(\ 2\pi(x-t)\ )\,, \tag{5.36}$$

so that the resulting metric is periodic and we can identify for instance the points -0.5 and 0.5 on the x axis. This allows to set up periodic boundary conditions in numerical simulations, so that the initial profile keeps turning around along the x direction. One can in this way test the long-term effect of these gauge perturbations.

According to the cross-comparison results [9], the linear regime (small amplitude, $A = 0.01$) poses no serious challenge to most numerical relativity codes (with the outstanding exception of the BSSN case, see for instance [10]). Allowing for this fact, we will focus in the medium and big amplitude cases ($A = 0.1$ and $A = 0.5$, respectively), in order to test the non-linear regime. Our simulations will take place in a 'numerical simulation channel' of $200 \times 8 \times 8$ grid points, with periodic boundary conditions along every axis. We are using here the first-order version of the Z4 system (4.19), (4.20), (4.21), (4.22), (4.23), (4.24), and (4.25) and the finite differences algorithm (5.29), with $s = 2$ (third order) and a dissipation parameter $b = 1/12$; this value will be justified later in the finite volume context.

The results of the numerical simulations are displayed in Fig. 5.1 for the H function (the γ_{xx} metric component). The left panel shows the medium

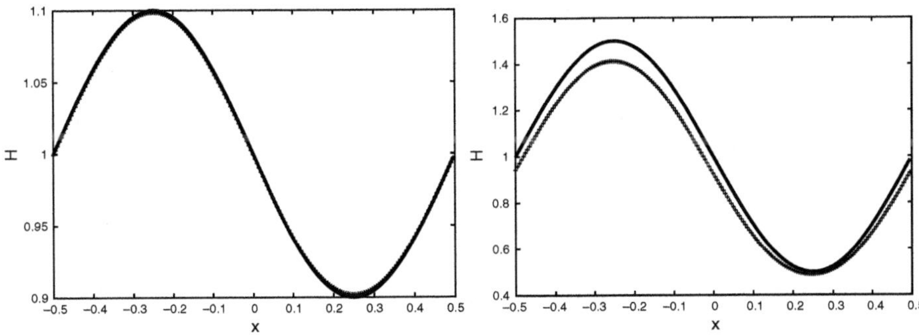

Fig. 5.1 Gauge waves simulation with periodic boundary conditions and sinusoidal initial data for the γ_{xx} metric component. The *left panel* corresponds to the medium amplitude case $A = 0.1$. After 1000 round trips, the evolved profile (cross marks) nearly overlaps the initial one (*continuous line*), which corresponds also with the exact solution. The right panel corresponds to the same simulation for the big amplitude case $A = 0.5$. We see the combination of a slight decrease in the mean value plus some amplitude damping.

amplitude case $A = 0.1$. Only a small amount of numerical dissipation is barely visible, even after 1000 round trips: the third-order-accurate finite-difference method gets rid of the dominant dispersion error. For comparison, let us recall that the corresponding BSSN simulation crashes before 100 round trips [11]. The right panel shows the same thing for the large amplitude case $A = 0.5$, well inside the non-linear regime. We see some amplitude damping after 1000 round trips, together with a slight decrease of the mean value of the lapse. These Z4 results are at the same quality level than the ones reported in the 'apples with apples' web page [9] for the flux-conservative generalized harmonic code Abigail, which is remarkable for a test running in strictly harmonic coordinates.

5.2 Finite volume methods

The finite difference methods presented in the previous sections assume the validity of the Taylor development of the physical solution. Only in this case we can talk about accuracy order in a rigorous sense. This is physically sound only when dealing with highly smooth solutions, like the ones verifying Einstein's field equations, except for the particular cases consisting in two or more matched spacetime domains (composite solutions). The matching conditions require just the continuity of the metric and its first derivatives: no consistent Taylor development is possible at the matching hypersurface, and finite-difference methods can get problems there as a result.

The situation is much worse in genuine non-linear fields, like in fluid dynamics, where shock discontinuities can arise even from very smooth initial data. This poses a stronger challenge, both at the continuum and at the discrete level. Computational fluid dynamics (CFD) is mainly based on finite volume methods, which can deal with these generalized (weak) solutions at the prize of a higher computational cost. We will see first some theoretical developments that justify the arising of those weak solutions and then a variety of numerical approaches that are currently being used in CFD.

5.2.1 Systems of balance laws

We have seen in Sect. 1.2.3 how Einstein's field equations can be interpreted as a system of balance laws. In the previous chapter, we have written the evolution equations in first-order form, like the ones appearing in the fluid dynamics domain. Now we are in a position to take full advantage of this analogy, both from the theoretical and from the practical point of view.

The principal part (4.26), (4.27), (4.28), (4.29), (4.30), (4.31), and (4.32) of the first-order Z4 system is in flux-conservative form. The same is true for the KST system (4.49), (4.50), and (4.51) and the generalized harmonic one. This means that the time evolution of the array $\mathbf{u}$ of dynamical fields can be written in the form

$$\partial_t \ \mathbf{u} + \partial_k \ \mathbf{F}^k = \cdots \,, \tag{5.37}$$

where the flux terms $\mathbf{F}^k$ depend algebraically on the fields, but not on their derivatives:

$$\mathbf{F}^k = \mathbf{F}^k(\mathbf{u}, x^j, t) \,. \tag{5.38}$$

If one takes into account the full system, not just the principal part, one gets the full balance law form:

$$\partial_t \ \mathbf{u} + \partial_k \ \mathbf{F}^k = \mathbf{S} \,, \tag{5.39}$$

where the source terms $\mathbf{S}$ do not contain any derivative,

$$\mathbf{S} = \mathbf{S}(\mathbf{u}, x^j, t) \,, \tag{5.40}$$

so that they do not contribute to the principal part.

The terms 'fluxes' and 'sources' come from the hydrodynamical analogous of the system (5.39). We can integrate term by term the differential system (5.39) over a given domain V in coordinate space in order to get the integral form of the balance law:

$$\partial_t \ [\int \mathbf{u} \ dV \ ] + \oint \mathbf{F}^k \ dS_k = \int \mathbf{S} \ dV \,, \tag{5.41}$$

where we have applied Gauss theorem and S_k stands for the boundary surface element along the coordinate direction x^k. The hydrodynamical analogy suggests to interpret $\mathbf{u}$ as a sort of density so that the rate of change of the integrated quantity $\bar{\mathbf{u}}$ defined as

$$\bar{\mathbf{u}} \equiv \int \mathbf{u} \, dV \tag{5.42}$$

depends on the integrated effect of the sources inside V

$$\bar{\mathbf{S}} \equiv \int \mathbf{S} \, dV \tag{5.43}$$

and on the fluxes across every boundary S_k of V, namely

$$\oint \mathbf{F}^k \, dS_k \,. \tag{5.44}$$

The balance law form (5.39) is specially suited for the MoL discretization, as described in Sect. 5.1.2. This is because in the method of lines there is a clear-cut separation between space and time discretization. As a consequence, the source terms contribute in a trivial way to the space discretization. The non-trivial contribution comes just from the flux-conservative part (5.37).

The integral version (5.41) provides then a useful way of getting a finite volume discretization of the first-order system (5.39). The space discretization can be obtained directly from

$$\partial_t \, \bar{\mathbf{u}} + \oint \mathbf{F}^k \, dS_k = \bar{\mathbf{S}} \,, \tag{5.45}$$

so that the evaluation of partial space derivatives has been replaced by that of surface integrals of the flux terms.

The finite volume mesh can be obtained as the dual of the finite differences grid, so that the elementary cells are centered on every grid node and their interfaces correspond to the intermediate points between neighbor nodes, as is displayed in Fig. 5.2 (only two space dimensions are shown for clarity).

Alternatively, a finite differences discretization can be directly obtained from the differential version (5.39). The resulting semi-discrete system would be given by

$$\begin{aligned} \partial_t \, \mathbf{u} = \mathbf{S} &- \frac{1}{\Delta x} \, [\mathbf{F}^x_{i+1/2} - \mathbf{F}^x_{i-1/2}] - \frac{1}{\Delta y} \, [\mathbf{F}^y_{j+1/2} - \mathbf{F}^y_{j-1/2}] \\ &- \frac{1}{\Delta z} \, [\mathbf{F}^z_{k+1/2} - \mathbf{F}^z_{k-1/2}] \,. \end{aligned} \tag{5.46}$$

Every way of computing the interface fluxes in terms of the values of the fields at the grid nodes will lead to a specific numerical algorithm.

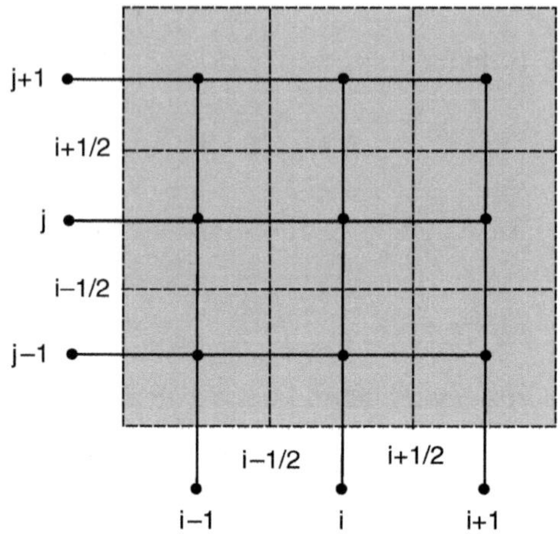

Fig. 5.2 The finite volume grid consists in a mesh of elementary cells, which can be interpreted as the dual of the finite differences grid (*solid lines*). The elementary cells, shown as *gray squares*, are centered around the corresponding nodes, shown here as *small dots*. The flux terms are evaluated at the interfaces (*dotted lines*), which are placed halfway between neighbor nodes.

5.2.2 Weak solutions

As stated in Sect. 1.2.3, there are composite solutions in which the metric coefficients can have piecewise continuous first derivatives. In first-order systems, these first derivatives are included in the array $\mathbf{u}$ of basic dynamical quantities (4.6). This means that the partial derivatives in the differential version (5.39) of the evolution system are not well defined.

To remedy this, one can complete the integration process leading from (5.39) to (5.45) by integrating also in time, that is,

$$\bar{\mathbf{u}}(\Delta t) - \bar{\mathbf{u}}(0) + \oint \bar{\mathbf{F}}^k \, dS_k = \int_0^{\Delta t} \bar{\mathbf{S}} \, dt \,, \tag{5.47}$$

where we have noted

$$\bar{\mathbf{F}}^k = \int_0^{\Delta t} \mathbf{F}^k \, dt \,, \tag{5.48}$$

so that no partial derivative appears in this integral version of the evolution system. Every solution of the differential version (5.39) is a smooth solution of the integral version (5.47). Conversely, there are solutions of the integral version (5.47) (weak solutions) which are not smooth.

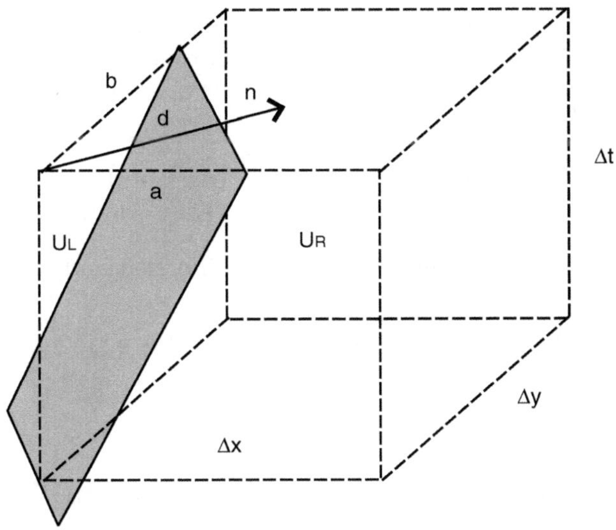

Fig. 5.3 An elementary cell is built in order to analyze the propagation of a discontinuity wavefront (*in gray*), which is entering by the *lower left corner* and propagating along the direction $\boldsymbol{n}$ (only two space dimensions are shown for clarity). The displacements of the discontinuity points along the x and y axes are labeled by a and b, respectively, whereas the wavefront displacement is labeled by d. The dynamical fields are supposed to be piecewise constant, with values u_R and u_L ahead and behind the wavefront, respectively.

The differential version (5.39) does not hold for weak solutions, unless we interpret it in terms of distributions. In this case, the piecewise continuous first derivatives can be described by using step functions on the surface of discontinuity. Derivatives across this surface, like the ones appearing on the left-hand side of (5.39), lead to singular Dirac delta terms in the equations. Cancellation of these singular terms will fix the propagation speeds of the discontinuities wavefronts, as we will see below.

We will go here one step further, by computing explicitly these propagation speeds. We will consider then a piecewise continuous solution in which the first derivatives are discontinuous across a wavefront surface propagating with coordinate speed v (see Fig. 5.3 for details). As we are interested only in the singular terms, we will make the following simplifications:

- The metric coefficients are supposed to be continuous.
- The source terms will be ignored, because they contain at most piecewise continuous contributions, but no Dirac delta singularities.
- The remaining dynamical fields in (4.6), corresponding to metric first derivatives, will be taken to be piecewise constant. The values ahead the wavefront are noted u_R, whereas the values behind it will be noted u_L.

This model problem is the general relativistic analogous of the well-known Riemann problem in fluid dynamics [12].

With these simplifying assumptions, we can easily evaluate the integral balance law (5.47) for the elementary cell shown in Fig. 5.3:

- For the density terms, we have

$$\bar{\mathbf{u}}(\Delta t) - \bar{\mathbf{u}}(0) = -\frac{a\,b}{2}\,(\mathbf{u}_R - \mathbf{u}_L)\,. \tag{5.49}$$

- The flux terms balance gives in turn (two directions only)

$$\oint \bar{\mathbf{F}}^k\; dS_k = \frac{\Delta t}{2}\;[\;b\,(\mathbf{F}^x_R - \mathbf{F}^x_L) + a\,(\mathbf{F}^y_R - \mathbf{F}^y_L)\;]\,. \tag{5.50}$$

- The final balance can be written then as

$$\frac{d}{a}\,(\mathbf{F}^x_R - \mathbf{F}^x_L) + \frac{d}{b}\,(\mathbf{F}^y_R - \mathbf{F}^y_L) = v\;(\mathbf{u}_R - \mathbf{u}_L)\,, \tag{5.51}$$

  where $v = d/\Delta t$ is the wavefront propagation speed.

It is easy to see from Fig. 5.3 that the coefficients in the left-hand side of (5.51) are precisely the components of the unit normal $\boldsymbol{n}$ to the wavefront. This means that we can put our final balance into its final form, valid for the generic three-dimensional case:

$$n_k\,(\mathbf{F}^k_R - \mathbf{F}^k_L) = v\;(\mathbf{u}_R - \mathbf{u}_L)\,. \tag{5.52}$$

These jump conditions are the general relativistic analogous of the well-known Rankine–Hugoniot conditions of fluid dynamics [12]. These equations govern the propagation of shocks or any other kind of discontinuities across the fluid. In the general relativity case, they provide in addition matching conditions for composite metrics.

Note that, for non-linear flux terms, the shock propagation speed v, as derived from (5.52), does not coincide with the characteristic speed λ. Physical shocks are formed when characteristic lines cross, so that we have

$$\lambda_L > v > \lambda_R\,. \tag{5.53}$$

When the information propagates along characteristic lines, this crossing produces a multivalued solution which is physically realized as a discontinuous wavefront. In the CFD language, the 'entropy condition' (5.53) means that the shock is supersonic as seen from behind, but subsonic when seen from the rear. We will see some examples of this genuine non-linear behavior at the end of this chapter.

On the contrary, Einstein's field equations are quasilinear. This means that we can express the flux terms in the form

$$\mathbf{F}^k = \mathbf{A}^k\, \mathbf{u}\,, \tag{5.54}$$

where the matrices $\mathbf{A}^k$ are continuous across the wavefronts because they depend on the metric coefficients, but not on their derivatives. Therefore, the jump conditions (5.52) can be written as a linear eigenvalue problem, namely

$$(\mathbf{A}^n - v\,\mathbf{I})\;(\mathbf{u}_+ - \mathbf{u}_-) = 0\,, \tag{5.55}$$

where the matrix $\mathbf{A}^n = n_k \mathbf{A}^k$ is the characteristic matrix along the direction $\boldsymbol{n}$. The eigenvectors of the characteristic matrix correspond to combinations of the dynamical fields that may have discontinuity surfaces with normal $\boldsymbol{n}$. The corresponding eigenvalues (characteristic speeds) provide the physical propagation speeds, so that

$$\lambda_L = v = \lambda_R\,. \tag{5.56}$$

These are called 'contact discontinuities,' and we have seen that these are the only ones allowed in quasilinear flux-conservative systems.

5.2.3 Flux formulae

The very existence of discontinuous solutions is a challenge for numerical algorithms. Let us go back to the flux-conservative semi-discrete system (5.46). The key point is to compute the interface fluxes $F_{j+1/2}$. A simple algorithm is obtained by using a plain average for the interface fluxes, namely

$$F^x_{i+1/2} \;=\; (F^x_i + F^x_{i+1})/2\,, \tag{5.57}$$

and the same way for the remaining directions. This amounts to discretize the flux derivative in (5.46) as

$$D^0_x\, F^x_i\,, \tag{5.58}$$

so that the resulting method is second-order accurate in space.

But if we want to allow for weak solutions, plain averages are not always the best option. Moreover, we can talk about the order of accuracy just in the formal sense, as the eventual discontinuities break down any Taylor development argument. We rather need a more general 'flux formula'

$$F^x_{i+1/2} = f(u_L, u_R)\,, \tag{5.59}$$

where u_L (resp. u_R) can be either u_i (resp. u_{i+1}) or any better estimation of the values on both sides of the cell interface.

5.2.3.1 Riemann-solver methods

An inspiring idea was proposed in the pioneering work of Godunov [13]. The point is to consider the numerical jump between the left and right predictions in (5.59) as a true physical discontinuity. One can obtain the physical solution to that 'Riemann problem' by enforcing the 'Rankine–Hugoniot' matching conditions (5.52) and then getting the interface flux from the physical solution. In this way, the flux formula (5.59) becomes a 'Riemann solver,' because it requires the exact solution of the corresponding Riemann problem.

The explicit form of this exact solution is known in some important CFD cases. But in the generic case one must recourse to approximate Riemann solvers, obtained usually from the linearized version of the equations (see for instance [14]). The main lines of the procedure for a generic strongly hyperbolic system would be as follows (we are computing here the F^x components):

- Decompose the system into its characteristic components (4.66). In the linearized case, we will get an advection equation for every eigenfield:

$$\partial_t \; w + \partial_x \; [\lambda \; w] = 0 \,. \tag{5.60}$$

- Take the upwind values of the characteristic fields at the interface, namely

$$w_{i+1/2} = \left\{ \begin{array}{ll} w_L & (\lambda > 0) \\ w_R & (\lambda < 0) \end{array} \right. . \tag{5.61}$$

- Recover the values of the dynamical fields at the interface $u_{i+1/2}$ from the characteristic ones and then take

$$F^x_{i+1/2} = F^x(u_{i+1/2}) \,. \tag{5.62}$$

This approximate Riemann-solver approach is actually a complicated process. Note for instance that the diagonalization is done at the grid points, but the recovery of the original fields is done at the cell interfaces. In the generic case, neither the diagonalization matrices nor the eigenvalues are constant, so this requires extra assumptions or averaging prescriptions. Moreover, the use of an approximate Riemann solver may introduce artificial (unphysical) shocks, where the entropy condition (5.53) is violated.

Riemann-solver methods have been common practice in CFD since decades. They were adapted to the numerical relativity context in early times [15], for dealing with the spherically symmetric (1D) black hole case. They are still considered as an option in relativistic hydrodynamics codes (see for instance [16]), but their use in 3D black hole simulations has been limited by the computational cost of performing the characteristic decomposition of the evolution system at every single interface.

5.2.3.2 Centered methods

More recently, much simpler alternatives have been proposed, which require just the knowledge of the characteristic speeds, not the full characteristic decomposition. They can be applied then directly to the primitive dynamical fields **u** and are called 'Centered methods,' in contrast to the upwind-biased prescriptions (5.61). Some of these methods have already been implemented in relativistic hydrodynamics codes [17].

Maybe the simplest choice is the local Lax–Friedrichs (LLF) flux formula [18]

$$f(u_L\,,\;u_R) = \frac{1}{2}\;[\;F_L + F_R + \lambda_M\;(u_L - u_R)\;]\;, \tag{5.63}$$

where the coefficient λ_M is an upper-bound estimate of the spectral radius of the dynamical system (the absolute value of the biggest characteristic speed), namely

$$\lambda_M = \max(\;|\lambda_L|\,,|\lambda_R|\;)\;, \tag{5.64}$$

λ being the values of the characteristic speeds.

The centered formula (5.63) can be understood as the plain average (5.57) plus some additional λ-terms. In the simplest case, when the left and right values are given by the nearest-neighbor estimates, the semi-discrete scheme corresponding to (5.63) can be written in the simple form as

$$\partial_t\,u_j = -D^0_x\,F_j + 1/2\,\Delta x\,D^+_x[\,\lambda_M\,D^-_x\,u_j\,]\,, \tag{5.65}$$

which corresponds to the lowest order ($s = 1$) finite-difference formula (5.29) with $b = 1/2$. We see then that the supplementary terms in (5.63) play the role of a numerical dissipation. In this sense, a much more dissipative choice would be

$$\lambda_M = \frac{\Delta x}{\Delta t}\;, \tag{5.66}$$

which corresponds to (a generalization of) the original Lax–Friedrichs algorithm [3].

A not-so-simple alternative to (5.63) is given by the HLL formula [19]:

$$f(u_L\,,\;u_R) = \frac{1}{2}\,[\,F_L{+}F_R{+}\frac{\lambda_+ + \lambda_-}{\lambda_+ - \lambda_-}\,(F_L{-}F_R){+}\frac{\lambda_+ - \lambda_-}{2}\,(u_L{-}u_R)\,]\,, \tag{5.67}$$

where we have noted

$$\lambda_+ = \max(\lambda_L, \lambda_R, 0) \qquad \lambda_- = \min(\lambda_L, \lambda_R, 0). \tag{5.68}$$

The HLL formula is currently used in relativistic hydrodynamics simulations [17], as it is less dissipative than the LLF one.

5.2.4 High-resolution methods

The flux formulae presented in the previous subsection require the knowledge of some left and right predictions for the dynamical field values at every cell interface. In the finite volume approach, however, we just have average values $\bar{\mathbf{u}}$ over the whole cell. Obtaining interface values from these averages requires to assume some specific profile for the dynamical fields $\mathbf{u}$.

The simplest assumption is that of a piecewise-constant profile, as shown in Fig. 5.4. Note that a discontinuity appears at every interface, where the values $\mathbf{u}_L$ are obtained from the previous cell and the values $\mathbf{u}_R$ are obtained instead from the next cell. To be more specific, at the $i + 1/2$ interface one gets

$$\mathbf{u}_L = \mathbf{u}_i \qquad \mathbf{u}_R = \mathbf{u}_{i+1}. \tag{5.69}$$

These one-sided predictions provide the simplest (lowest order) predictions for any specific choice of flux formula.

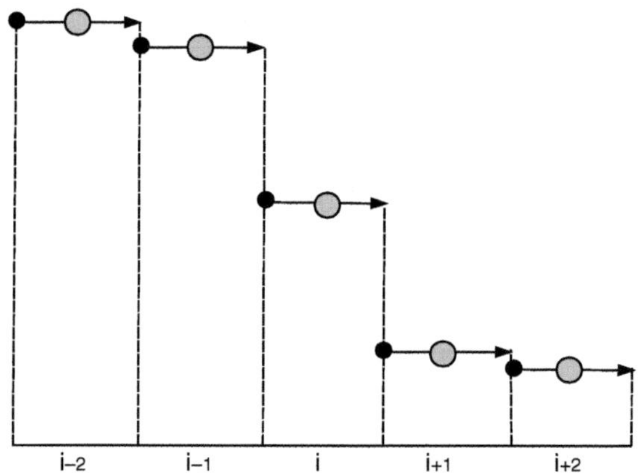

Fig. 5.4 Flat (piecewise constant) reconstruction from the average values of the dynamical fields. The averages coincide with the values at the grid nodes (*gray circles*). Numerical discontinuities appear at every interface (*dotted lines*) between the left and right values (*arrows* and *dots*, respectively).

In order to verify the properties of the resulting algorithms, we will consider also the simplest LLF formula (5.63). Moreover, as (the principal part of) any strongly hyperbolic system amounts to a set of advection equations, we will consider here just one of these advection equations with a generic characteristic speed v. The corresponding flux will be given then by

$$F(u) = v\ u \tag{5.70}$$

(we will consider for the moment only one space direction).

Let us apply to this simple case the piecewise-constant reconstruction (5.69). The corresponding discretization can be obtained by replacing the prescriptions (5.63) and (5.69) into the general expression (5.46). The result is the linear three-point algorithm:

$$u_i^{n+1} = u_i^n + \frac{\Delta t}{2\Delta x}\,[\,(\lambda^M_{i+1/2} - v_{i+1})\,u^n_{i+1} + (\lambda^M_{i-1/2} + v_{i-1})\,u^n_{i-1} - (\lambda^M_{i+1/2} + \lambda^M_{i-1/2})\,u_i^n\,]\,. \tag{5.71}$$

Allowing for the fact that λ^M is chosen at every interface as the absolute value of the maximum speed on both sides, we can see that all the u^n coefficients are positive provided that the Courant stability condition

$$\lambda^M\,\frac{\Delta t}{\Delta x}\ \leq\ 1 \tag{5.72}$$

is satisfied everywhere. Note, however, that a more restrictive condition would be obtained in the three-dimensional case, where we must add up in (5.71) the contributions from every space direction.

The positivity of all the coefficients means that values u^{n+1} provided by the algorithm (5.71) for the next time level are just weighted averages of the values u^n for the current time level. The stability of the algorithm is ensured, as a blowup can never be obtained just by averaging the field values. Moreover, it is easy to show that the positivity property implies that a monotonic profile will remain monotonic, so spurious oscillations cannot appear ('monotonicity preserving' see for instance [12]). The positivity property implies then some strong form of stability. In the MoL approach, this strong stability is preserved by time-evolution algorithms with positive coefficients, like (5.22). That is the reason for the term 'strong-stability-preserving' (SSP) algorithms.

From the accuracy point of view, however, the algorithm (5.71) is disappointing: just first-order accurate. This is because of the crude (piecewise constant) reconstruction which leads to the interface predictions (5.69). 'High-resolution' methods can be built by assuming instead a piecewise-linear reconstruction of the interface values, as shown in Fig. 5.5), that is,

$$\mathbf{u}_L = \mathbf{u}_i + \frac{1}{4}\,(\mathbf{u}_{i+1} - \mathbf{u}_{i-1}) \tag{5.73}$$

$$\mathbf{u}_R = \mathbf{u}_{i+1} - \frac{1}{4}\,(\mathbf{u}_{i+2} - \mathbf{u}_i)\,, \tag{5.74}$$

which will ensure second-order spatial accuracy. There is a price to pay for this: as we can see in Fig. 5.5, the linear reconstruction does not preserve the monotonicity of the dynamical fields. This will cause spurious oscillations in the numerical results, as we will see later.

Note that this monotonicity breaking can occur only when the slope that is being used for the reconstruction of a given field u gets at least twice as

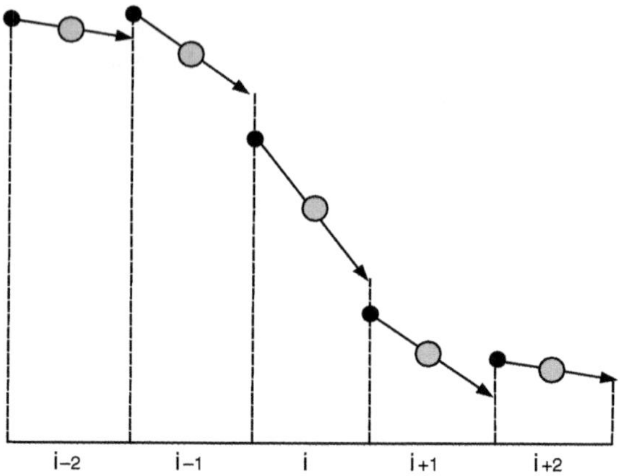

Fig. 5.5 Same as in the previous figure, but now for a piecewise linear reconstruction. Numerical discontinuities appear again at every interface between the *left* and *right* values. Note that the original function was monotonically decreasing: all the slopes are negative. In this linear reconstruction, however, both the *left* interface values (at $i - 3/2$) and the *right* interface ones (at $i + 3/2$) show local extreme values that break the monotonicity of the original function.

large as any of the left- or right-sided slopes:

$$\Delta_i^L \equiv (u_i - u_{i-1})\,, \qquad \Delta_i^R \equiv (u_{i+1} - u_i) \tag{5.75}$$

(see Fig. 5.5). Otherwise, the predictions at a given interface are always bounded between the neighbor node values so that monotonicity of the node values would imply that of the interface ones. In our case, we have chosen the average

$$\Delta_i^C \equiv (\Delta_i^L + \Delta_i^R)/2 = (u_{i+1} - u_{i-1})/2 \tag{5.76}$$

(centered slope).

A simple calculation shows that, in this case, the monotonicity breaking can occur only if one of the left and right slopes (5.75) at a given point is at least three times larger than the other. This gives a more precise sense to the 'steep gradient' notion in the centered slopes case. In CFD terms, this is a 'compression factor' of three, which is quite acceptable. It is unlikely that we find steeper profiles when dealing with smooth solutions, provided that we use a numerical grid with the required resolution. But this can be a problem in at least two scenarios:

- When one is dealing with piecewise continuous solutions, where the discontinuities appear independent of the space resolution. This is the case of weak solutions, currently arising in hydrodynamics and MHD simulations.
- When one is dealing with (smooth) solutions that get steep gradients at specific locations during evolution. This is actually the case in black hole

simulations when one uses singularity-avoidant slicing conditions, as we will see in the next chapter.

The natural way to remedy this is to enforce that both (left and right) interface predictions are in the interval limited by the corresponding left and right point values (interwinding requirement). A monotonicity-preserving algorithm can be obtained then by using the non-linear 'monotonic centered' (MC) slope reconstruction [20]

$$\Delta^{MC} \equiv \text{minmod}(\ 2\Delta^L\ ,\ \Delta^C\ ,\ 2\Delta^R\ )\,, \tag{5.77}$$

instead of the linear average (5.76). The minmod function is defined as follows:

- If all the arguments have the same sign, then it selects the one with smaller absolute value.
- If one of the arguments has different sign than the others, then it is zero.

In this way, the slopes are limited in order to avoid spurious oscillations. The rule is that interface values must lie between their neighbor node values (see Fig. 5.6). There is an extra amount of numerical dissipation at the extreme points (where space accuracy is reduced just to first order), but this is common to most of the 'High-resolution methods' that are currently used in CFD. The particular reconstruction we have presented here belongs to the class of 'slope-limiter' methods, as described in [12, 14]. When combined with a specific Riemann solver, it is known as MUSCL method [21] (for monotonic upstream centered limiter). The MUSCL method is currently implemented in relativistic hydrodynamics 3D simulations (see for instance [22]).

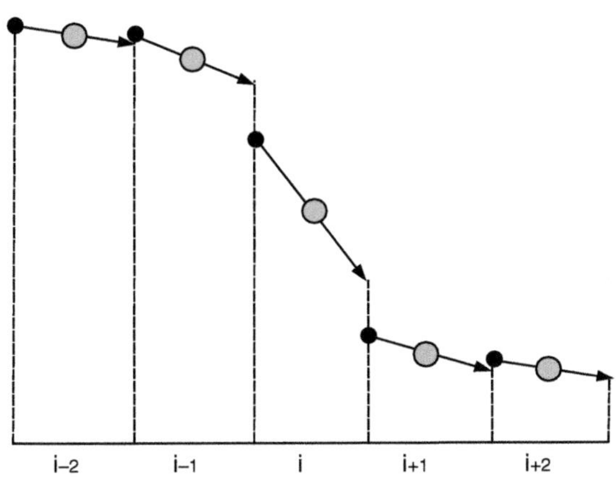

Fig. 5.6 Same as in the previous figure, but using the monotonic-centered reconstruction. Note that the interface values are bounded now between the neighbor nodes, so that monotonicity is preserved both for the left values (*arrows*) and for the right ones (*dots*) at every interface (*dotted lines*).

5.2.5 The modified flux approach

The 'high-resolution' methods that we have seen up to now are based on a linear reconstruction of the dynamical fields. In a uniform grid, this allows to identify the cell averages with the central (node) values, as we have done implicitly in the previous sections. The step to 'very high-resolution' methods, that is, increasing accuracy beyond second order in space, would require a non-linear cell reconstruction, like in 'piecewise parabolic' (PPM) methods [23]. Then, identifying averages with node values is no longer possible and the resulting schemes are more complicated.

An alternative approach is to compute directly the interface fluxes. The starting point can be any lowest order flux $h_{j+1/2}$, derived from one of the flux formulae described in Sect. 5.2.3. The idea is to compute the required flux $F_{j+1/2}$ by adding some correction terms

$$F_{j+1/2} = h_{j+1/2} + df_{j+1/2}\,, \tag{5.78}$$

which are devised for improving the accuracy of the algorithm (modified flux approach [24]). The flux corrections (5.78) can be obtained by combining one-sided contributions from the neighbor nodes, namely

$$df^{+}_{j+1/2} = F_{j+1} - h_{j+1/2}, \qquad df^{-}_{j+1/2} = h_{j+1/2} - F_j\,. \tag{5.79}$$

Among the different implementations of this approach in CFD, we will choose here that of Osher and Chakrabarthy [25, 26], given by

$$df_{j+1/2} = \sum_{k=-s+1}^{s-1} \left(c^s{}_k \; df^{-}_{j+k+1/2} + d^s{}_k \; df^{+}_{j+k+1/2} \right)\,, \tag{5.80}$$

where

$$d^s{}_k = \nu^s{}_k - (-1)^k\, b \begin{pmatrix} 2s-2 \\ k+s-1 \end{pmatrix}\,, \qquad c^s{}_k = -d^s{}_{-k} \tag{5.81}$$

(b is here an arbitrary parameter), and

$$\nu^s_0 = 1/2\,, \qquad \nu^s_k = -\nu^s{}_{-k} \qquad (k \neq 0) \tag{5.82}$$

$$\nu^s_{s-1} = (-1)^{s-1} \left[s \begin{pmatrix} 2s \\ s \end{pmatrix} \right]^{-1} \qquad (s > 1) \tag{5.83}$$

$$\nu^{s+1}_k = \nu^s_k + (-1)^k\, \frac{k}{s} \begin{pmatrix} 2s \\ s-k \end{pmatrix} \left[(s+1) \begin{pmatrix} 2s+2 \\ s+1 \end{pmatrix} \right]^{-1}. \tag{5.84}$$

For the simpler non-trivial cases we have (decreasing k order)

$$d^2_k = \left\{ b - \frac{1}{12}, \frac{1}{2} - 2\,b,\, b + \frac{1}{12} \right\} \tag{5.85}$$

$$d_k^3 = \left\{ \frac{1}{60} - b, 4\,b - \frac{7}{60}, \frac{1}{2} - 6\,b, 4\,b + \frac{7}{60}, -\frac{1}{60} - b \right\}. \tag{5.86}$$

We have seen in the previous section how the positivity property of lowest order flux formulae is lost in high-accuracy algorithms and how it can be restored by using 'slope limiters' of the form (5.77), at the price of losing accuracy where the limiters actually modify the slope values. In this modified flux context, this is done rather by limiting the flux corrections $df^{\pm}$. To be more specific, every flux correction contribution to (5.80) must be replaced by

$$\widetilde{df^{\pm}}^{k}_{j+k+1/2} = \text{minmod}\,(\,df^{\pm}_{j+k+1/2},\, c\,df^{\pm}_{j+1/2},\, c\,df^{\pm}_{j-1/2}\,)\,, \tag{5.87}$$

where c is some compression factor. A sufficient condition for positivity is given then by [7]

$$1 + d_{-1}^{\,s} - d_0^{\,s} + c \sum_{k\neq 0} \min\,(\,d_{k-1}^{\,s} - d_k^{\,s},\, 0\,) \;\geq\; 0\;, \tag{5.88}$$

plus the following restriction on the time step (1D case):

$$\lambda_j\, \frac{\Delta t}{\Delta x}\, [\; d_{-1}^{\,s} - d_0^{\,s} + c \sum_{k\neq 0} \max\,(\,d_{k-1}^{\,s} - d_k^{\,s},\, 0\,)\;] \;\leq\; 1/2\;, \tag{5.89}$$

where we have assumed $d_k^{\,s} = 0$ when $|k| \geq s$.

Allowing for (5.88), for the third-order-accurate algorithm ($s = 2$), the maximum compression factor is obtained for [7]

$$b = \frac{1}{12}\;, \qquad c_{\max} = 5\,, \tag{5.90}$$

whereas for the fifth-order-accurate case ($s = 2$), one gets the optimal value

$$b = \frac{2}{75}\;, \qquad c_{\max} = \frac{265}{83}\;, \tag{5.91}$$

which is still larger than the corresponding value of 3 for the MC slope limiter (5.77) in the (second-order) MUSCL scheme.

The compression factor (5.90) obtained for the optimal value of the dissipation parameter is surprisingly high. This means that the flux-correction limiters (5.87) would have no effect for smooth monotonic solutions, unless we get a poor resolution. This fact suggests the possibility of switching off those flux limiters in some scenarios, like spacetime evolution, in which weak solutions are not expected to arise.

We must remember here that any lowest order flux $h_{j+1/2}$ can still be selected among the ones described in Sect. 5.2.3. Let us choose for simplicity the local Lax–Friedrichs flux (5.63). The flux differences (5.79) in this case get the simple form

$$df^{\pm}_{j-1/2} = 1/2\ [\,F_j - F_{j-1} \,\pm\, \lambda_M\,(u_j - u_{j-1})\,]\ . \qquad (5.92)$$

In the 'unlimited' case, the corresponding development provides the flux-conservative form of the semi-discrete algorithm

$$\partial_t u_j \,=\, -C^{2s}F_j + (-1)^{s-1}\, b\,(\Delta x)^{2s-1} D_+^s\, D_-^{s-1}(\lambda_M\, D_- u_j)\ , \qquad (5.93)$$

which is $2s-1$ order accurate in space, with a stencil of $2s+1$ points. This is actually the algorithm (5.29) that we introduced in the finite-differences context. We will call these numerical schemes FDOC, for 'finite-difference Osher–Chakrabarthy'. We will use the third-order-accurate FDOC algorithm (5.93), with the optimal parameters (5.90) as a default for spacetime evolution and even as an option for matter evolution, as we will see in the next section.

Let us remark also that the choices (5.90) and (5.91) are optimal for a generic choice of the lowest order TVD flux. In the LLF case, however, it is clear that the spectral radius can be multiplied by a global magnifying factor $K > 1$. Allowing for the finite-difference form (5.93) of the unlimited version, magnifying λ amounts to magnify b, that is,

$$(b,\ K\,\lambda) \qquad \Leftrightarrow \qquad (K\,b,\ \lambda)\ . \qquad (5.94)$$

It follows that the values of the compression factor $b_{\max}$ obtained in the previous section must be interpreted just as lower-bound estimates. In particular, the equivalence (5.94) implies that any compression factor bound obtained for a particular value b_0 applies as well to all values $b > b_0$. This agrees with the interpretation of the second term in (5.93) as modeling numerical dissipation. On the other hand, this dissipation term is actually introducing the main truncation error. We will use then in what follows the b values in (5.90) and (5.91), which are still optimal in the sense that they provide the lower numerical error compatible with the highest lower bound for the compression parameter.

5.3 Simple CFD tests

We will present here some simple tests, currently used in the CFD domain. We will go from the simplest linear case up to the complex non-linear MHD systems. We will test in all cases both the finite differences FDOC algorithm and an implementation of the MUSCL type, with the LLF flux formula and the MC slope limiter. In this way, we can compare the accuracy of the FDOC scheme with the robustness of the MUSCL one, where the slope limiters get rid of any spurious oscillations at the price of some local accuracy loss.

5.3.1 Advection equation

Let us start by the scalar advection equation:

$$\partial_t u + \lambda\, \partial_x u = 0 \qquad (\lambda \text{ constant}). \tag{5.95}$$

This is the simplest linear case, but it allows to test the propagation of arbitrary initial profiles, containing jump discontinuities and corner points, departing from smoothness in many different ways. This is the case of the Balsara–Shu profile [27], which will be evolved with periodic boundary conditions.

Let us show first, in Fig. 5.7, the results of the MUSCL reconstruction method, with the MC slope limiter (5.77). In this simple case, all the reported flux formulae are equivalent. Note that no spurious oscillations appear, due to the positivity of the resulting scheme. To be more precise, we can define in this simple case the total variation (TV) of the solution as

$$TV(u) = \sum_j \,|\, u_j - u_{j-1}|\,. \tag{5.96}$$

In the case of systems, the total variation is defined as the sum of the total variation of the components. The MUSCL scheme is total-variation-diminishing (TVD), meaning that $TV(u)$ does not increase during numerical evolution. It is obvious that TVD schemes cannot develop spurious oscillations: monotonic initial data preserve their monotonicity during time

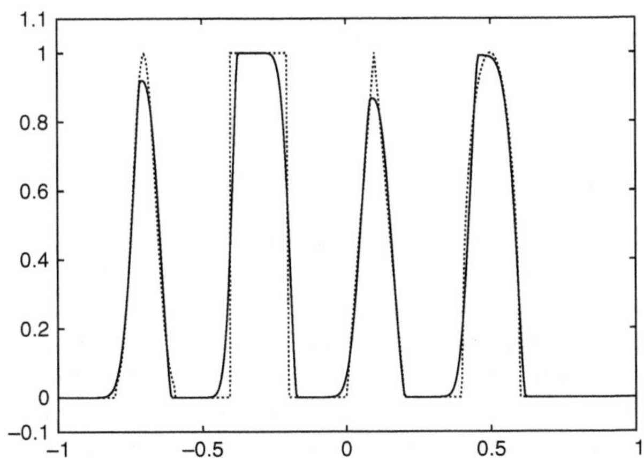

Fig. 5.7 Advection of the Balsara-Shu profile in a numerical mesh of 400 points. A MUSCL scheme is used with the MC slope limiter. The results are compared with the initial profile (*dotted line*) after a single round trip. No spurious oscillations appear, although the effect of numerical dissipation is clearly visible, both at the peaks and in the corners roundoff.

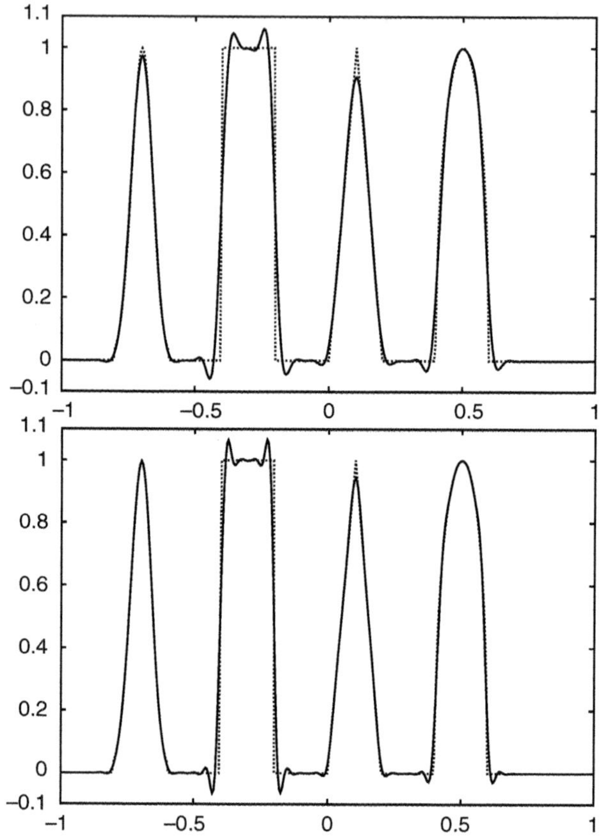

Fig. 5.8 Same as the previous figure, but now for using FDOC algorithms, with either third-order (*upper panel*) or fifth-order (*lower panel*) accuracy.

evolution. Moreover, the TVD property can be seen as a strong form of stability: any blowup of the numerical solution is excluded, as it would increase the total variation.

We show in Fig. 5.8 the corresponding numerical result for the FDOC algorithm, either for the third-order formula ($s = 2$, $b = 1/12$) or for the fifth-order one ($s = 3$, $b = 2/75$). The smooth regions are described correctly: even the height of the two regular maxima is not reduced too much by dissipation, as expected for an unlimited algorithm with at most fourth-order dissipation. There is a slight smearing of the jump slopes, as usual for contact discontinuities, which gets smaller with higher resolution.

Concerning monotonicity, it is clear that the total variation of the initial profile has increased by the riddles besides the corner points and, more visibly, near the jump discontinuities. We can see in the lower panel that one additional riddle appears at every side of the critical points, due to the larger (seven-point) stencil. On the other hand, we have checked that the total vari-

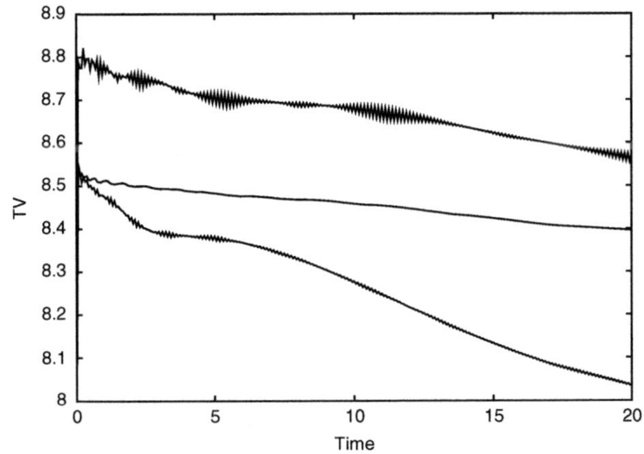

Fig. 5.9 Advection equation (FDOC case). Time evolution of the total variation. The horizontal axis corresponds to the exact solution: $TV(u) = 8$. From *top* to *bottom*: $s = 3$ scheme with 400 points, $s = 2$ scheme with 800 points, and $s = 2$ scheme with 400 points. After the initial increase, which depends on the selected method, TV tends to diminish. Increasing resolution just reduces the diminishing rate.

ation is bounded for this fixed time, independently of the space resolution or the time step size, that is,

$$TV(u) \leq B \,, \tag{5.97}$$

where the upper bound B is independent of the resolution, but could depend on the elapsed time.

This statement can be verified by plotting, as we do in Fig. 5.9, the time evolution of $TV(u)$ for the FDOC case. In all cases, a sudden initial increase is followed by a clear diminishing pattern. These numerical results indicate that the bound on the total variation is actually time independent, beyond the weaker TVB requirement.

Even if we were ready to relax the stronger TVD requirement, keeping the bound (5.97) is important from the theoretical point of view. One major advantage of total-variation-bounded (TVB) schemes is that the numerical solutions converge locally (in L^1 norm) to a weak solution of (the integral version of) the evolution equation. If an additional entropy condition is satisfied, then the proposed scheme is convergent (see for instance [12]).

5.3.2 Burgers equation

Burgers equation provides a simple example of a genuinely non-linear scalar equation:

$$\partial_t \, u + \frac{1}{2} \, \partial_x \, (u^2) = 0 \,. \tag{5.98}$$

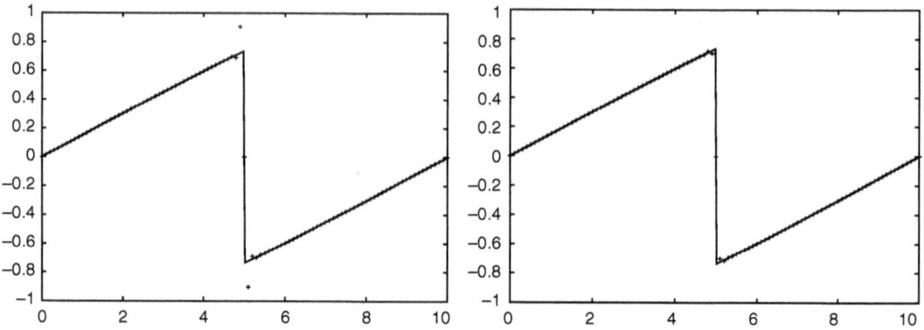

Fig. 5.10 Burgers equation: evolution of an initial sinus profile. The numerical solution (*point values*) is plotted versus the exact solution (*continuous line*), for the FDOC and the MUSCL schemes (*left* and *right panels*, respectively), with a 100 points resolution.

The exact solution of this differential equation can be given in parametric form as

$$u = h(p) \qquad p = x - ut\,, \tag{5.99}$$

where $h(x)$ is any initial profile. For instance, we can consider

$$h(x) = \sin(\,\frac{x\,\pi}{5}\,)\,, \tag{5.100}$$

which is a smooth, even analytical, profile. Note that the characteristic speed is u itself, so that characteristic lines converge at the origin. Allowing for (5.99), we see that the solution uniqueness is lost for $t > 10/\pi$, meaning that characteristic lines start crossing.

A true shock develops as a result from these smooth initial data, which propagates, allowing for the Rankine–Hugoniot condition (5.52), with the shock speed

$$v = (u_R + u_L)/2\,. \tag{5.101}$$

For the symmetric initial data (5.100), this gives $v = 0$, meaning that the shock will stay fixed at the origin. We plot in Fig. 5.10 the numerical solution values versus (the principal branch of) the exact solution, at a time where the shock has fully developed. We compare the FDOC and the MUSCL results (left and right panels, respectively). In the MUSCL case, the TVD property is preserved at the price of some extra dissipation near the shock, where the slope limiters are doing their job. On the contrary, in the FDOC case, we can see again some spurious oscillations which affect mainly the points directly connected with the shock.

These conclusions are fully confirmed by a second simulation, obtained by adding a constant term to the previous initial profile, that is,

$$h(x) = \frac{1}{2} + \sin(\frac{x\,\pi}{5})\,, \tag{5.102}$$

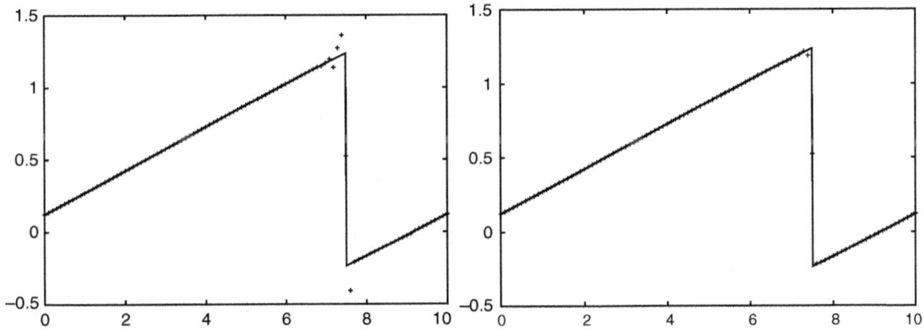

Fig. 5.11 Same as in the previous figure, but now for a moving sinus initial profile. The numerical solution (*point values*) is plotted versus the exact solution (*continuous line*) for 100 points resolution and for the FDOC and the MUSCL schemes (*left* and *right panels*, respectively).

with periodic boundary conditions. We can see in Fig. 5.11 that a shock again develops, but it does no longer stand fixed: it propagates to the right. The plots shown correspond to $t = 7$.

Note that the Burgers differential equation (5.10) can also be written as

$$\partial_t \, u + u \; \partial_x \, (u) = 0 \,, \tag{5.103}$$

suggesting that the 'sonic points' ($u = 0$) should stay fixed during evolution. This is not actually the case, as the shock is propagating through all the domain with a non-vanishing speed given by (5.101). This is because weak solutions are not solutions of the differential equation. Weak solutions obey instead the integral form of the equations, which can be derived from the flux-conservative form (5.98), rather that from (5.103).

5.3.3 Euler equations: Sod test

Euler equations for fluid dynamics are a convenient arena for testing the proposed schemes beyond the scalar case. In the ideal gas case, we can check the numerical results against well-known exact solutions containing shocks, contact discontinuities, and rarefaction waves.

We will deal first with the classical Sod shock-tube test [28] with a standard 200 points resolution. The tube is filled with an ideal gas, initially at rest, with two different homogeneous states separated by a wall. The left and right initial states are characterized by the following values of density and pressure:

$$\rho_L = p_L = 1.0 \qquad \rho_R = p_R = 0.125 \,. \tag{5.104}$$

The local characteristic speed is given by

$$\lambda = v \pm c_s \qquad c_s = \sqrt{\gamma \frac{p}{\rho}} \,, \tag{5.105}$$

where v is the fluid speed and c_s is the sound speed (we take the adiabatic index $\gamma = 1.4$, corresponding to air).

Let us remove the wall at $t = 0$ and follow the fluid evolution. We plot in Fig. 5.12 the gas density profile at a time when there is still some fluid on both sides unaffected by the wall removal. At the left of the initial wall position (at $x = 0.5$), we see a rarefaction wave moving with negative speed. The term rarefaction means that the density is diminishing as the wave propagates. The term 'rarefaction fan' is also currently used, because characteristic lines are diverging there like the sticks of a folding fan. As a consequence, no shock appears on the left-hand side. Note that the fluid is moving rightward there, as it follows from the velocity plots in Fig. 5.13.

The situation is just the opposite on the right-hand side, where characteristic lines converge and a shock is generated as a result. This right-propagating shock is specially visible as the only discontinuity in the speed profile (see Fig. 5.13). An intermediate constant-speed stage is formed, and we can see a second discontinuity in the density plots. This is a contact discontinuity, which propagates with the positive sound speed in (5.105): an 'acoustic wave.'

Concerning the numerical results, we see that both the rarefaction wave and the shock are perfectly resolved, whereas the contact discontinuity is smeared out in both cases. Resolving contact discontinuities is always a challenge for numerical schemes based on centered flux formulae: dealing with the exact solution (Riemann solver) would of course improve this, at a much higher computational cost. In the FDOC case, we see some overshots just besides the shock, specially visible in the speed profile, where the jump is much higher. Concerning the MUSCL scheme, the overshots are avoided at the price of some extra numerical dissipation, which can be seen in the rarefaction wave and at the contact discontinuity (density profile).

5.3.4 MHD equations: Orszag–Tang vortex

As a simple multi-dimensional example, let us consider here the Orszag-Tang vortex problem [29]. This is a well-known model problem for testing the transition to supersonic magnetohydrodynamical (MHD) turbulence and has become a common test of numerical MHD codes in two dimensions.

A barotropic fluid (with adiabatic index $\gamma = 5/3$) is considered in a doubly periodic domain, with uniform density ρ and pressure p. A velocity vortex given by $\mathbf{v} = (-\sin y, \sin x)$ is superimposed on the initial data with a magnetic field $\mathbf{B} = (-\sin y, \sin 2x)$, describing magnetic islands with half the

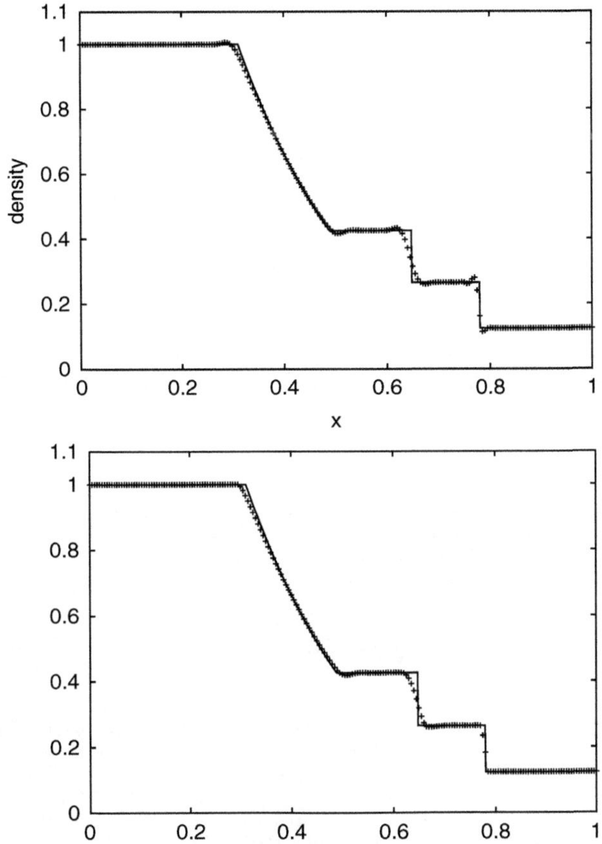

Fig. 5.12 Sod shock-tube problem. Density profile for the FDOC and the MUSCL schemes (*upper* and *lower panels*, respectively). From *left* to *right*, we see the *left* initial state, a rarefaction wave, two constant-density intermediate states connected by a contact discontinuity, a shock, and the *right* initial state.

horizontal wavelength of the velocity roll. As a result, the magnetic field and the flow velocity differ in their modal structures along one spatial direction.

In Fig. 5.14 the temperature, $T = p/\rho$, is represented at a given time instant ($t = 3.14$). The figure clearly shows how the dynamics is an intricate interplay of shock formation and collision. The FDOC numerical scheme, with $s = 2$ and $b = 1/12$, seems to handle the Orszag-Tang problem quite well.

In Fig. 5.15 we plot the results for the same problem using a MUSCL-type second-order scheme built from an approximate Riemann solver of the Roe-type and the MC limiter. The results with both methods are qualitatively very similar. The overshots in the FDOC case can be seen as very thin lines (white or dark) profiling the temperature contrasts. The greater dissipation in the MUSCL case can be seen instead as smoothing these contrasts.

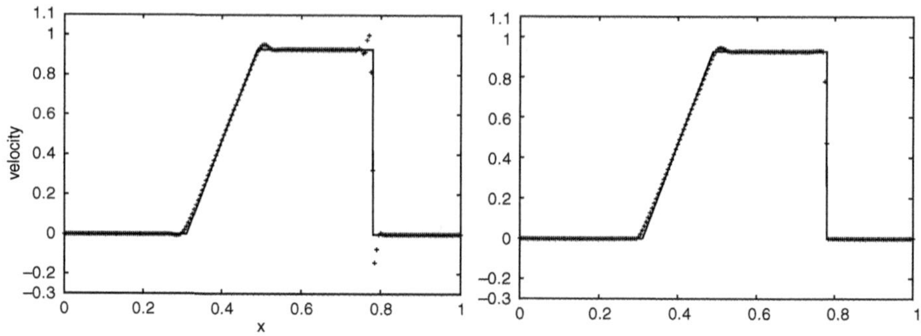

Fig. 5.13 Same as the previous figure, but for the speed profiles. Sod shock-tube problem. The FDOC and the MUSCL results are shown (*left* and *right panels*, respectively). Note that the fluid is moving everywhere to the *right*, toward the *lower* density region. Note also that the fluid speed is constant across the contact discontinuity, which is not visible here.

Fig. 5.14 Temperature at $t = 3.14$ in the Orszag-Tang vortex test problem. In this simulation, the third-order FDOC scheme has been used with a numerical grid of 200×200 mesh points.

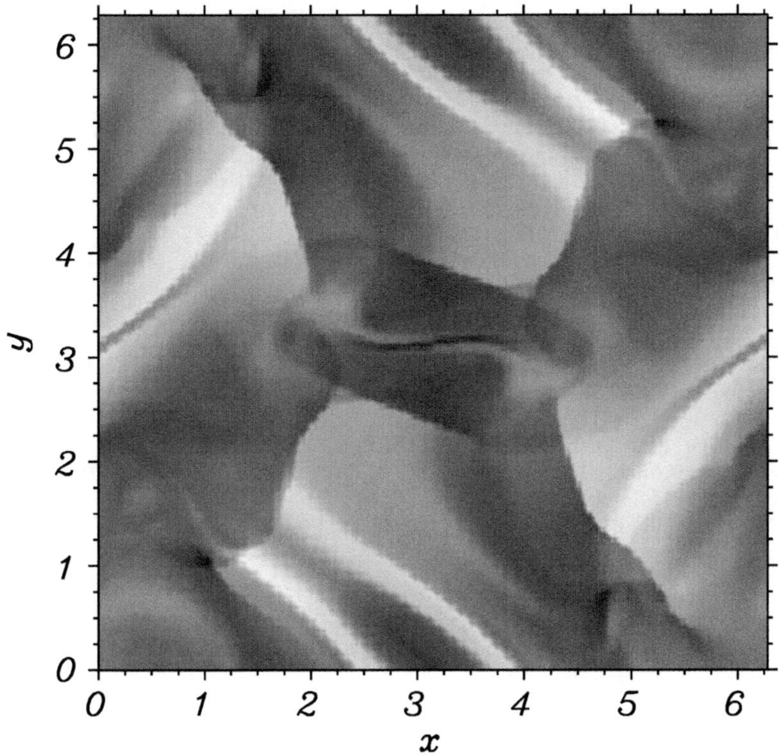

Fig. 5.15 Same as in the previous figure, but now for a second-order MUSCL scheme built from the Roe-type solver and the MC limiter.

References

1. P. D. Lax, B. Wendroff, Commun. Pure Appl Math. **13**, 217–237 (1960).
2. R. W. MacCormack, AIAA Paper 69–354 (1969).
3. W.H. Press, B.P. Flannery, S.A. Teukolsky and W.T. Veterling, *Numerical Recipes*, Cambridge University Press, Cambridge(1989).
4. O. A. Liskovets, J. Diff. Eq. **I**, 1308–1323 (1965).
5. C.-W. Shu and S. Osher, J. Comp. Phys. **77**, 439–471 (1988).
6. B. Gustafson, H.O. Kreiss and J. Oliger, *Time dependent problems and difference methods*, Wiley, New York (1995).
7. C. Bona, C. Bona-Casas and J. Terrades, J. Comp. Phys. **228**, 2266 (2009).
8. M. Alcubierre et al, Class. Quantum. Grav. **21**, 589 (2004).
9. http://www.appleswithapples.org/TestMethods/Tests/tests.html
10. K. Kiuchi and H-A. Shinkai, Phys. Rev. D**77**, 044010 (2008).
11. Y. Zlochower, J. G. Baker, M. Campanelli and C. O. Lousto, Phys. Rev. D**72**, 024021 (2005).
12. R. J. LeVeque, *Numerical methods for conservation laws*, Birkhauser, Basel, (1992).
13. S. K. Godunov, Math. Sbornik **47** 271–306 (1959), translated US Joint Publ. Res. Service, JPRS **7226**, (1969).

14. E. Toro, *Riemann Solvers and Numerical Methods for Fluid Dynamics*, Springer, Berlin Heidelberg (1997).
15. C. Bona and J. Massó, Phys. Rev. Lett. **68**, 1097 (1992).
16. http://www.whiskycode.org
17. M. A. Aloy, J. A. Pons and J. M. Ibáñez, Computer Physics Commun. **120** 115 (1999).
18. V. V. Rusanov, J. Comput. Mat. Phys. USSR **1** 267 (1961).
19. A. Harten, P. D. Lax and B. van Leer, SIAM Rev., **25**, 35 (1983).
20. B. van Leer, J. Comput. Phys. **23**, 276 (1977).
21. B. van Leer, J. Comput. Phys. **32**, 101 (1979).
22. F. Banyuls, J. A. Font, J. M. Ibáñez, J. M. Martí, and J. A. Miralles, Astrophys. J. **476**, 221 (1997).
23. P. Colella and P. R. Woodward, J. Comput. Phys. **54**, 174 (1984).
24. A. Harten, J. Comput. Phys. **49**, 357–393 (1983).
25. S. Osher and S. Chakravarthy, ICASE Report 84–44 (1984), *IMA Volumes in Mathematics and its Applications*, Vol 2, pp. 229–274 Springer-Verlag New York (1986).
26. C.-W. Shu, Math. Comput. **49**, 105–121 (1987).
27. D. S. Balsara and Chi-Wang Shu, J. Comput. Phys. **160**, 405–452 (2000).
28. G. A. Sod, J. Comput. Phys. **27**, 1-31 (1978).
29. S. A. Orszag and C. M. Tang, J. Fluid Mech. **90**, 129–143 (1979).

Chapter 6
Black Hole Simulations

The essential ingredients of a numerical relativity code are the evolution formalism and the numerical scheme. We have already discussed a couple of well-tested evolution formalisms: the generalized harmonic one and Z4, from which BSSN can be derived by symmetry breaking. We have also presented a robust, cost-efficient, finite-difference scheme (FDOC), which is able to evolve smooth solutions, and a more sophisticated alternative (MUSCL), suitable for weak solutions.

Black hole simulations, however, still represent some challenge for numerical relativity codes. This is because of the nature of the geometry that we are trying to model. Singularities can arise even at the initial data. Also, specific mechanisms must be implemented in order to prevent the time lines of the reference observers to fall inside the hole or, from the reciprocal point of view, the black hole to grow up to the boundary of our numerical domain. Singularity-avoidant gauge conditions are of some help, but at the price of severely deforming the time slices geometry. In some cases, this goes beyond the limits allowed by our numerical resolution. The main alternative, dynamically excising the interior region, is also challenging. We will discuss these topics in this chapter at the basic level, although the reader is referred to the current research reviews for more elaborate developments (see for instance [1, 2]).

6.1 Black Hole initial data

In order to start a numerical simulation, we must provide the initial values of every dynamical field (we assume here that we are using a first-order evolution system). As we are trying to get a true Einstein solution, these initial data cannot be prescribed arbitrarily: the initial values of $(\gamma_{ij},\, K_{ij})$ must verify the energy and momentum constraints:

Bona, C. et al.: *Black Hole Simulations*. Lect. Notes Phys. **783**, 143–170 (2009)
DOI 10.1007/978-3-642-01164-1_6

$$\mathcal{E} \equiv \frac{1}{2}\,[\,tr\,R + (tr\,K)^2 - tr(K^2)\,] - \tau = 0 \tag{6.1}$$

$$\mathcal{M}_i \equiv \nabla_j\,(K_i{}^j - trK\,\delta_i{}^j) - S_i = 0\,. \tag{6.2}$$

This is true even if we are using the Z4 formalism, where we must also impose

$$\Theta = 0\,, \qquad Z_i = 0 \tag{6.3}$$

on the initial slice. The algebraic conditions (6.3) are not first integrals of the Z4 evolution system (4.19), (4.20), (4.21), (4.22), (4.23), (4.24), and (4.25). Therefore, the vanishing of the first time derivative of (6.3) must also be imposed on the initial slice and, allowing for (4.24) and (4.25) this amounts precisely to the energy and momentum constraints (6.1) and (6.2). The second-order subsidiary equation (3.74) ensures that no further conditions are required on the initial data in order to recover a true Einstein solution.

The easiest (but tricky) way of solving the initial data constraints (6.1) and (6.2) is just to take a t = constant slice of a known exact solution of Einstein's field equations. The simplest black hole initial data can be obtained from the Schwarzschild line element. Let us start for simplicity by considering the normal coordinates case (zero shift). The 'Schwarzschild coordinates' expression (1.54) can be replaced by the 'isotropic coordinates' version:

$$-\left(\frac{\rho - M/2}{\rho + M/2}\right)^2\, dt^2 + \left(1 + \frac{M}{2\rho}\right)^4\, \delta_{ij}\; dx^i dx^j\,, \tag{6.4}$$

which is better suited for the 3D case.

We note a number of features that can then be extrapolated from (6.4) to other (simple) sets of black hole initial data:

- It is a **vacuum** solution. This is good for testing just the geometrical sector of the numerical codes, independent of the hydrodynamical sector, where matter quantities should be computed from their own evolution equations.
- It is initially at rest (**time-symmetric** initial data), that is,

$$K_{ij}\,|_{t=0} \;=\; 0\,. \tag{6.5}$$

  This means that the momentum constraint (6.2) automatically holds true. The energy constraint (6.1) gets then the simpler form

$$tr\,R = 0. \tag{6.6}$$

- The space metric γ_{ij} is **conformally flat** (conformal to the Euclidean metric). This can be used in order to further simplify (6.6), as we will see in what follows.

- The space metric contains a **singularity**. This is the origin of the main complications we will face in this chapter, and this is why black hole simulations deserve a special treatment. In the Schwarzschild case, the singularity is located at the origin
$$\rho = 0\,, \tag{6.7}$$
where the space metric in (6.4) diverges. Note that the line element (6.4) has also a coordinate singularity (vanishing of the lapse), which corresponds to the apparent horizon (6.10), as we will see below.
- There is an **apparent horizon**. It can be defined as the two-surface where outgoing light rays have zero expansion, that is,
$$\theta = \nabla_k\, n^k + K_{ij}\, n^i n^j - trK = 0\,, \tag{6.8}$$
where n is here the outgoing unit normal to this two-surface (the wavefront). Quite surprisingly, only the dynamical field values at a given time slice are needed in order to locate the apparent horizon (which is a local, slicing-dependent feature). It follows from (6.8) that, for time-symmetric initial data, apparent horizons are also **minimal surfaces** (surfaces of minimal area), that is,
$$\nabla_k n^k = 0\,. \tag{6.9}$$
In the Schwarzschild case, it is natural to consider spherical wavefronts. Allowing for (6.4) and (6.9), the apparent horizon on the initial slice is the sphere given by
$$\rho = M/2\,, \tag{6.10}$$
which corresponds to $r = 2M$ in Schwarzschild coordinates.

In order to provide a geometrical interpretation of these features, we will analyze in more detail the Schwarzschild space metric in (6.4). Let us consider for instance the geometry of a plane passing through the origin. The 2D metric for this plane can be written in polar coordinates as

$$d\sigma^2 = \left(1 + \frac{M}{2\rho}\right)^4\; [\; d\rho^2 + \rho^2 d\varphi^2 \;]\,. \tag{6.11}$$

The line element (6.11) can also be considered as the metric of some 2D surface in the 3D Euclidean space. In this way, the geometry of the original plane in the Schwarzschild (curved) space is 'embedded' in Euclidean space, which is more familiar to everyone.

We have plotted in Fig. 6.1 the surface with metric (6.11) in flat space. Quite surprisingly, the singularity at $\rho = 0$ arises from the fact that the 'center' ($\rho = 0$) really corresponds to a spherical surface at space infinity. There is a minimal surface (corresponding to the minimal circle in the figure) at $\rho = M/2$, so that lower values of ρ lead back to circles of greater area. This minimal surface (the 'throat' in Fig. 6.1) coincides also with the location of the apparent horizon in the initial (time-symmetric) slice, as it follows by comparing (6.8) with (6.9). The overall image is a sort of 'wormhole,' where the interior geometry is just a replica of the exterior one.

6.1.1 Conformal metric decomposition

The fact that the space metric in (6.4), namely

$$\left(1+\frac{M}{2\rho}\right)^4 \delta_{ij}\,, \tag{6.12}$$

is conformally flat suggests to consider the conformal decomposition of a generic 3D metric:

$$\gamma_{ij} = e^{2U}\ \hat{\gamma}_{ij}\,, \tag{6.13}$$

where $\hat{\gamma}_{ij}$ is a given conformal metric (the Euclidean one in the Schwarzschild case).

The connection coefficients of the original and the conformal space metric are related by

$$\Gamma^k_{ij} = \widehat{\Gamma}^k_{ij} + \delta^k{}_i\, U_j + \delta^k{}_j\, U_i - U^k \hat{\gamma}_{ij}\,. \tag{6.14}$$

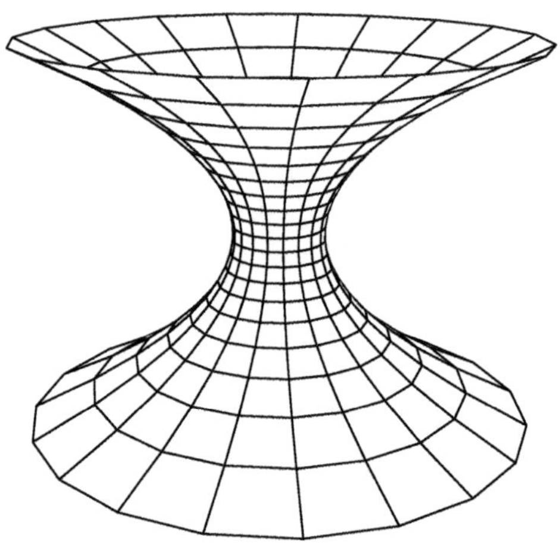

Fig. 6.1 Embedding of the Schwarzschild space metric as a ordinary surface in Euclidean spacetime. It is the revolution surface obtained from a horizontal parabola with vertex at the Schwarzschild radius ($\rho = M/2$ in isotropic coordinates). *Radial lines* beginning at infinity in the *upper side* come to the *center*, but reach a minimal circle at $\rho = M/2$ (the locus of the parabola vertices, seen as a 'throat' in the plot) and then go back by the *lower side* again to infinity. The singularity comes precisely from the counterintuitive fact that $\rho = 0$ is not actually the *center point*, but it corresponds instead to a spherical surface placed at spatial infinity.

It follows that the corresponding Ricci tensors verify

$$R_{ij} = \widehat{R}_{ij} - \widehat{\nabla}_i U_j + U_i\, U_j - (\widehat{\nabla}_k U^k + U_k U^k)\, \hat{\gamma}_{ij}\,, \tag{6.15}$$

where the conformal metric is used in all index contractions. The corresponding relationship between both Ricci scalars is then

$$tr\, R = \widehat{tr}\, \widehat{R} - 4\, \widehat{\nabla}_k U^k - 2\, U_k U^k\,. \tag{6.16}$$

When the conformal metric $\hat{\gamma}_{ij}$ is flat, that is,

$$\widehat{R}_{ij} = 0\,, \tag{6.17}$$

it is convenient to express the conformal factor in (6.13) in the form

$$e^{2U} = \Psi^4\,, \tag{6.18}$$

so that the vacuum time-symmetric energy and momentum constraints reduce to the Laplace equation on Ψ [3]:

$$\widehat{\triangle}\, \Psi = 0\,. \tag{6.19}$$

By a suitable choice of the space coordinates on the initial slice, one can always assume that the flat conformal metric takes the Euclidean form, namely

$$\hat{\gamma}_{ij} = \delta_{ij}\,. \tag{6.20}$$

The spherically symmetric solutions of (6.19) can then be obtained from

$$\partial_\rho\, [\, \rho^2\, \Psi_\rho\,] = 0\,, \tag{6.21}$$

so that one gets the Schwarzschild conformal factor in (6.12) as a result.

6.1.2 Singular initial data: punctured black holes

We will consider here a class of initial data which is common to many black hole simulations:

- Time-symmetric, conformally flat initial slice, that is,

$$K_{ij}\,|_{t=0} \;=\; 0\,, \qquad \gamma_{ij}\,|_{t=0} \;=\; \Psi^4 \delta_{ij}\,, \tag{6.22}$$

  where Ψ is a solution of (6.19).
- Gauss initial gauge

$$\alpha\,|_{t=0} \;=\; 1\,, \qquad \beta^i\,|_{t=0} \;=\; 0\,. \tag{6.23}$$

- No extra fields on the initial slice. In the Z4 formalism, for instance,

$$\Theta\,|_{t=0} \;=\; 0\,, \qquad Z_i\,|_{t=0} \;=\; 0\,. \tag{6.24}$$

The problem of getting consistent initial data leading to true Einstein's solutions is then reduced to finding solutions of the 3D Laplace equation (6.19) in the conformal Euclidean space.

This is exactly the same equation as that of the electrostatic potential in vacuum. We know that any non-trivial electrostatic potential is generated by some charge distribution. This means that the solutions Ψ of the vacuum Laplace equation will usually contain singularities, corresponding to the location of point charges acting as sources. We have already seen this in the Schwarzschild case: the 'potential'

$$\Psi = 1 + \frac{M}{2\rho} \tag{6.25}$$

can be interpreted in electrostatic terms as a trivial (constant) contribution plus a Coulomb term at the origin.

It follows from these considerations that the Schwarzschild case is a good representative of black hole initial data, in the sense that 'puncture' singularities appear in the initial metric through the conformal factor Ψ [4, 5] (the term 'puncture' is actually suggested by the embedded geometry diagram in Fig. 6.1). Dealing with these singularities in numerical simulations is then the very first challenge.

The 'punctured black holes' approach was originally devised in the framework of the BSSN formalism. The initial (singular) conformal factor

$$e^{2U} = \Psi^4\,|_{t=0} \tag{6.26}$$

was entered analytically into the code, so that only the (non-singular) dynamical deviations were actually computed numerically [6]. The conformal metric $\hat{\gamma}_{ij}$ was then evolved by starting from the regular initial data

$$\hat{\gamma}_{ij}\,|_{t=0} \;=\; \delta_{ij}\,. \tag{6.27}$$

In summary, the singular contributions were computed analytically and the regular ones numerically.

This clever approach has an inherent drawback: The singular term contributions are always there, as they are included analytically in the modified evolution equations. This fact prevented for years getting long-term simulations of orbiting black holes. The 2005 breakthrough in this field was actually made by a completely different approach: excision in the generalized harmonic formalism [7].

Just after this, the 'moving punctures' approach (a modification of the original one) allowed the same kind of success for BSSN simulations without excision [8, 9]. The idea was just to evolve everything numerically, even the singular terms. The numerical grid was arranged so that the initial singularities did not sit on any of the numerical nodes. Artificial dissipation was in charge for smoothing the numerical solution, so that the singularities 'leak out' through the finite difference mesh after a number of iterations.

Of course, it is difficult to get a rigorous concept of convergence in this way (the true continuum solution is singular). One can interpret that a 'numerical migration' has taken place from a singular solution to a regular one, to which the numerical values are actually converging. Although this approach has shown to be extremely successful from the practical point of view, some doubts can arise about the physics that has just 'leaked out' through the mesh. This is why other approaches, dealing with regular initial data, deserve to be considered.

6.1.3 Regular initial data

One can wonder whether it is possible at all to start a black hole simulation with regular initial data. The idea is to take advantage of the 'one-way membrane' paradigm: no physically meaningful information can get out from the horizon in a (classical) black hole. This means that one can modify the geometry of the interior region without affecting the exterior one, provided that it is done in a consistent way. One can use this idea to get rid of the initial singularities, as we will see below.

To be more precise, a number of comments are in order:

- The one-way membrane is not the apparent horizon (a local feature), but the event horizon, which can be defined as the boundary of the region which has no causal effect at infinity. This is a non-local feature (requires information from all future slices) although it can be computed numerically [10, 11]. However, as far as we know, the apparent horizon is interior to the event horizon, so that a safer strategy is to restrict any change to the interior of the apparent horizon, which can be easily located at any time slice.
- Gauge speed can be higher than light speed. This means that gauge modes can cross the horizon in both directions. However, this is not against the one-way membrane idea, because gauge conditions are not part of the field equations nor carry any physically relevant information: one can choose them in quite an arbitrary way. Gauge modifications inside the horizon can then be regarded just as one more way of determining the coordinate conditions.
- The stability of the numerical code requires that the numerical speed (3.31) be higher than the physical speed. It follows that numerical errors prop-

agate faster than light and, then, they can also cross the horizon in both directions. Any change whatsoever in the interior region must then be consistent and accurate if one wants the get a physically sound solution for the black hole exterior region.

6.1.3.1 Stuffed black holes

Regular initial data can easily be obtained from 'composite' space geometries. The idea comes once more from the electrostatic analogy suggested by (6.19), which we will write here in the non-vacuum case as

$$\widehat{\triangle}\,\Psi = -\frac{1}{4}\,\tau\,\Psi^5\,. \tag{6.28}$$

It is well known that the external (Coulomb) field of a spherical charge distribution is the same as the one generated by a point charge. In the same way, the exterior Schwarzschild geometry is the same as the one generated by a interior matter distribution with the same total mass.

The key point is to match the Schwarzschild exterior space metric with a (non-singular) interior space metric corresponding to some matter distribution ('stuffed' black holes). Although it has been done previously for the dust case [12], we prefer to consider here the scalar field case, as it has shown more robustness in long-term simulations [7, 13]. The field equations read in this case

$$R_{\mu\nu} = 8\pi\,(\partial_\mu \Phi)\,(\partial_\nu \Phi)\,, \tag{6.29}$$

so that the 3+1 decomposition of the energy–momentum tensor is given by

$$\tau = 1/2\,({\Phi_n}^2 + \gamma^{kl}\Phi_k\,\Phi_l)\,, \qquad S_i = \Phi_n\,\Phi_i\,, \tag{6.30}$$

$$S_{ij} = \Phi_i\,\Phi_j + 1/2\,({\Phi_n}^2 - \gamma^{kl}\Phi_k\,\Phi_l)\,\gamma_{ij}\,, \tag{6.31}$$

where Φ_n stands for the normal time derivative

$$(\partial_t - \beta^k\,\partial_k)\,\Phi = -\alpha\,\Phi_n\,, \tag{6.32}$$

and Φ_i for the partial space derivative of Φ.

The stress–energy conservation amounts to the evolution equation for the scalar field, which is just the scalar wave equation. In the 3+1 language, it translates into the flux-conservative form:

$$\partial_t\,[\sqrt{\gamma}\,\Phi_n\,] + \partial_k\,[\sqrt{\gamma}\,(-\beta^k\Phi_n + \alpha\,\gamma^{kj}\Phi_j)\,] = 0\,. \tag{6.33}$$

A fully first-order system may be obtained by considering the space derivatives Φ_i as independent dynamical fields, along the lines discussed in Chap. 4.

Concerning the initial data, we must solve the energy–momentum constraints (6.1) and (6.2). In the time-symmetric case, this amounts to

$$R = 16\pi\,\tau\,, \qquad S_i = \Phi_n\,\Phi_i = 0\,. \tag{6.34}$$

The momentum constraint can be satisfied by taking Φ (and then Φ_i) to be zero everywhere on the initial time slice.

We will keep considering the conformally flat line element (6.4), so that the energy constraint can still be written as (6.28). But we will assume a variable mass profile $m = m(\rho)$, with a constant mass value $m = M$ for the black hole exterior, so that (6.28) will be satisfied with $\tau = 0$ there. In the interior region, however, the energy constraint will translate instead into the equation

$$m'' = -2\pi\rho\;(\Phi_n)^2\;(1+\frac{m}{2\rho})^5\;, \tag{6.35}$$

which can be interpreted as providing the initial Φ_n value for any convex ($m'' \leq 0$) mass profile. Of course, some regularity conditions both at the center and at the matching point ($\rho = M/2$) must be assumed. Allowing for (6.35), we must have

$$\begin{array}{lll} & m = m'' = 0 & (\rho = 0) \\ m = M, & m' = m'' = 0 & (\rho = M/2)\;. \end{array}$$

For instance, a simple profile verifying these conditions is given by

$$m(\rho) = 4\rho - 4/M[\,\rho^2 + (M/2\pi)^2\;\sin^2(2\pi\rho/M)\,]\;. \tag{6.36}$$

Note that, allowing for (6.35), the matching conditions ensure just the continuity of Φ_n, not its smoothness. This can cause some numerical error, as we are evolving Φ_n through the differential equation (6.33) with finite difference methods. If this is a problem, we can demand the vanishing of additional derivatives of the mass function $m(\rho)$, both at the origin and at the matching point.

6.1.3.2 Free Black Holes

An even simpler alternative is possible in the framework of the Z4 formalism (but see [14] for a recent implementation of the same idea in the BSSN context). In this case, the energy and momentum constraint violations are represented by the supplementary dynamical fields:

$$Z_\mu \Leftrightarrow (\Theta,\;Z_k)\,. \tag{6.37}$$

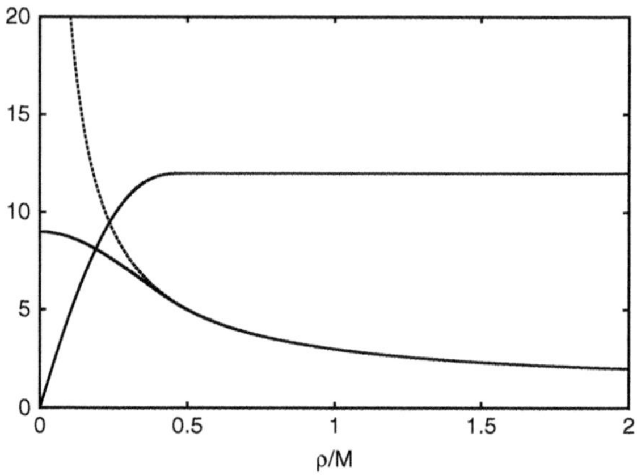

Fig. 6.2 Plot of a conformal factor Ψ providing time-symmetric conformally flat initial data for a free Schwarzschild black hole. Values on the x axis correspond to the isotropic radial coordinate ρ, measured in units of M. The singular expression (6.25) in the interior region (*dotted line*) is replaced by a regular one (*continuous line*), corresponding to (6.36). The mass profile is also shown in the figure (at a bigger scale). Both expressions coincide in the exterior (constant mass) region. Note that the matching is very smooth at the apparent horizon ($\rho = M/2$).

When non-zero values of these fields can be evolved without problems, we can conclude that the Z4 formalism is tolerant to energy and momentum constraint violations.

Moreover, we know that both Θ and Z_k can be computed from light-cone eigenfields, with light speed as characteristic speed. It follows that non-zero values of these supplementary fields in the initial data would propagate along light cones. In the region exterior to the black hole, the outgoing part will reach the outer boundary and leave out the computational domain. The incoming part will instead cross the horizon, entering the interior region. As nothing traveling with light speed can get out of the black hole interior, non-zero values of Θ and Z_k will cumulate there.

Playing again with the idea that the interior region has no causal physical influence on the exterior one, we can devise a simple way of obtaining regular initial data without either using excision or stuffing:

- Use the physical initial data (the ones verifying the energy and momentum constraints) in the exterior region.
- Forget about the energy and momentum constraints in the interior region (free black holes). Just take there any regular and smooth extension of the exterior geometry.

This approach is different from that of the matching problem (finding a suitable interior solution for a given exterior three-metric), which was that of the previous section. Here, one just extends smoothly the exterior geometry without imposing any constraint on the interior part. Only regularity conditions must be required.

Note, however, that removing the matter content without changing the metric can be interpreted as adding some matter counterterms which cancel out the physical ones. Of course, these extra matter terms are unphysical (carrying negative energy density, for instance). The resulting simulations can show some counterintuitive behavior as a result: gravitational collapse being stopped locally at the interior, and so on. This approach must be taken then with a lot of caution.

6.1.3.3 Black Hole excision

Another way of getting black hole regular initial data is to excise any region, interior to some apparent horizons, that contains singular points [15]. This means that we are setting inner boundaries on our computational grid and suitable boundary conditions must be provided there. The problem with inner boundary conditions is that they should hold in the strong-field high-speed region. This is in contrast with outer boundary conditions, which could in principle be placed in the weak-field zone, making things much easier.

One interesting possibility is to place the inner boundaries at minimal surfaces (the throat in Fig. 6.1). In the wormhole geometries we are considering, the interior region is just a replica of the exterior one. This is not so surprising if one remembers that the Laplace equation (6.19) is the one currently used in electrostatics, where the method of images provides precisely this kind of mirror-like solutions. In the Schwarzschild case, the inversion transformation

$$\rho \qquad \longleftrightarrow \qquad \frac{M^2}{4\rho} \tag{6.38}$$

leaves the spacetime metric (6.4) invariant.

The discrete mirror-like symmetry (6.38) allows one to set up 'virtual points' near the inner boundary by using just physical information from the grid nodes, without any extra input or assumption. After all, every point inside the throat is identical to another one outside, both related by (6.38). In the spherically symmetric case, this provides a nice way of excising the singular region which has been used successfully in (1D) black hole simulations [16–18].

In the generic 3D case, however, inner boundaries are not usually placed at the throats. The most common choice is to place the inner boundaries close to the (expanding) apparent horizon [19]. In contrast with the fixed throat location, apparent horizons are usually expanding with time. Inner boundaries are then forced to move during evolution, departing from their initial location at the throat. This makes the whole issue much more complicated.

An alternative approach to the excision boundary problem is the use of a different representation of the Schwarzschild line element. Instead of the isotropic form (6.4) of the line element, we can get back the original radial Schwarzschild coordinate r, but with a different time slicing, namely

$$\overline{t} = t + 2M \, ln \, |\frac{r}{2M} - 1| \,, \tag{6.39}$$

so that the metric gets the Eddington–Finkelstein form:

$$g_{\mu\nu} = \eta_{\mu\nu} + \frac{2M}{r} \, l_\mu l_\nu, \qquad l_\mu = (1, \frac{x}{r}, \frac{y}{r}, \frac{z}{r}) \,. \tag{6.40}$$

The corresponding 3+1 decomposition is given by

$$g_{ij} = \delta_{ij} + \frac{2M}{r} \, \frac{x_i}{r} \frac{x_j}{r}, \tag{6.41}$$

$$\alpha^2 = \left(1 + \frac{2M}{r}\right)^{-1}, \qquad \beta_i = \frac{2M}{r} \, \frac{x_i}{r} \,, \tag{6.42}$$

so that the extrinsic curvature reads

$$\alpha \, K_{ij} = \frac{2M}{r^2} \, [\, \delta_{ij} - \left(2 + \frac{M}{r}\right) \frac{x_i}{r} \frac{x_j}{r} \,] \,. \tag{6.43}$$

Initial data can be constructed directly from the previous expressions. The energy–momentum constraints (6.1) and (6.2) will be automatically satisfied, as far as (6.40) is an exact vacuum solution. The only singular point will be the origin. But note that the causal character of time lines in (6.40) is given by the sign of

$$g_{00} = -1 + \frac{2M}{r} \,, \tag{6.44}$$

so that they become spacelike in the black hole interior ($r < 2M$ in these coordinates). This means that all characteristic speeds will have the same (inward) sign at the interior.

One can take advantage of this fact by setting up an excision boundary fully inside the horizon and devising one-sided numerical algorithms which do not require any information from the excised points (see for instance [20]). The stability of this 'excision without boundary' numerical approach relies on the fact that all the characteristic speeds have actually the same sign, pointing to the excised region, outside the computational domain. Of course, one must switch to the standard centered algorithms at the points outside the horizon, where the propagation speeds get different signs.

This excision-without-boundary approach is actually the method of choice in black hole simulations based on the generalized harmonic formalism [7, 20, 21]. In some cases, it is combined with a scalar field stuffing of the initial data [22]. Initial data for multiple black holes can also be constructed, even if we no longer have an exact solution at our disposal (see for instance [23]).

6.2 Dynamical time slicing

One can wonder why to choose black hole regular initial data if singularities will appear anyway during time evolution. This is a good point. In the stuffed black hole case, for instance, the interior region is filled with matter, which will collapse by its own gravity to a singularity in a finite amount of proper time.

Collapse singularities arise when two nearby, but different, time lines (the world lines of two different Eulerian observers) meet at the very same point. Then, the proper distance between these time lines, which are labeled with different space coordinates, vanishes. This means that the space metric is no longer invertible there (the space volume element $\sqrt{\gamma}$ vanishes). Of course, this would be fatal for a numerical simulation, which will crash when trying to compute the (divergent) components of the inverse space metric.

This behavior is generic in black hole spacetimes. Remember that the apparent horizon is defined as the surface for which the expansion of a outgoing congruence of light rays vanishes (6.8). This expansion is then negative inside the apparent horizon, meaning that even outgoing light rays are actually collapsing. As far as the time slices are spacelike hypersurfaces, their normal lines (the world lines of the Eulerian observers) are timelike, that is, interior to the local light cones. It follows that the expansion (2.29) of the Eulerian observers must also be negative, that is,

$$\theta < 0 \;\Leftrightarrow\; trK > 0 \tag{6.45}$$

inside the apparent horizon, so that the normal lines are actually collapsing.

6.2.1 Singularity avoidance

We have seen in Sect. 2.1.2 how coordinate time (the time label that our computer is using when proceeding from one time slice to the next) is related with proper time through the lapse function α. The idea is to slow down the evolution, by choosing smaller values of α, in the regions that are going to collapse, whereas keep constant values of α in the outer regions, where information about the collapse can actually be recovered. The sequence of time slices would then cover most of the outer region, while keeping safely away from the singularity which is going to form inside the black hole.

Let us be more precise. Let us choose the normal time line corresponding to any fixed point x_0 on the initial slice (normal coordinates). Let us suppose now that this time line is going to meet a collapse singularity in a finite amount of proper time, that is,

$$\sqrt{\gamma} \rightarrow 0 \qquad \text{for} \qquad \tau \rightarrow \tau_S\,. \tag{6.46}$$

The coordinate time elapsed in the process is then given by

$$\Delta t = \int_0^{\tau_S} \frac{d\tau'}{\alpha}\,. \tag{6.47}$$

A necessary condition for the singularity to be avoided is that (6.47) be an improper integral, because the result is the coordinate time at which the singularity will actually occur. This means that the lapse function α should 'collapse' to zero, that is,

$$\alpha \rightarrow 0 \qquad \text{for} \qquad \tau \rightarrow \tau_L \leq \tau_S\,. \tag{6.48}$$

Otherwise, it is clear that the singularity will be attained in a finite amount of coordinate time.

We will assume for the moment that the lapse collapses precisely at the singular point, that is,

$$\tau_L = \tau_S \tag{6.49}$$

(the case $\tau_L < \tau_S$ will be considered in the next section). In this case, singularity avoidance is achieved if and only if the improper integral (6.47) diverges, so that the singularity is not attained in a finite amount of coordinate time. We can put this condition in an equivalent differential form: the improper integral (6.47) will diverge if and only if the proper time derivative of the lapse does have a finite value at the singularity, namely

$$\partial_\tau\, \alpha \;|_{\tau=\tau_S} \;=\; \lim \left(\frac{\alpha}{\tau - \tau_S}\right) \;<\; \infty\,. \tag{6.50}$$

We can check this condition for the class of generalized harmonic slicing conditions (3.89) that we are considering here,

$$\partial_\tau\, \alpha = -f\alpha\, tr\, K \tag{6.51}$$

(remember that $\Theta = 0$ for true Einstein's solutions), which we know leads to strongly hyperbolic evolution systems for $f > 0$. It follows that singularity avoidance will be achieved in our case if and only if

$$f\alpha\, tr\, K \;|_{\tau=\tau_S} \;<\; \infty\,. \tag{6.52}$$

We can still refine a little bit more our analysis by assuming that

$$\partial_\tau \sqrt{\gamma} \; < \; \infty \,. \tag{6.53}$$

This means that we are dealing with just ordinary collapse singularities ('focusing singularities' in [24]) and not with stronger ones in which the space volume element could vanish suddenly, at an infinite proper time rate. It follows that, for generalized harmonic slicing, singularity avoidance is achieved if and only if either there is a limit surface or

$$\frac{f\alpha}{\sqrt{\gamma}} \; < \; \infty \,. \tag{6.54}$$

Simple examples of singularity avoidance are provided by the choice $f = n$ (constant). In this case condition (6.51) can easily be integrated to give

$$\alpha/\alpha_0 = (\gamma/\gamma_0)^{n/2} \,, \tag{6.55}$$

so that the lapse is collapsing precisely at the singular point, where the space volume element is vanishing (no limit surface appears). Allowing for (6.54), the singularity will not be reached in a finite amount of coordinate time if and only if

$$n \geq 1 \,. \tag{6.56}$$

The case $n = 1$ corresponds to the original harmonic slicing condition, which verifies (6.56) marginally. This is why it usually requires the excision of the region close to the singularity in black hole numerical simulations.

6.2.2 Limit surfaces

A popular choice in black hole numerical simulations is given by the generalized '1+log' slicing condition $f = n/\alpha$. Condition (6.51) can easily be integrated again to give

$$\alpha - \alpha_0 = \frac{n}{2} \; ln(\gamma/\gamma_0) \,, \tag{6.57}$$

which justifies the '1+log' name (the initial lapse is usually taken to be one). Note that the lapse collapses here even before the singular point is reached, at a surface defined by

$$\gamma = \gamma_L \equiv \gamma_0 \; \exp{(-\frac{2\alpha_0}{n})} \; > \; 0 \,. \tag{6.58}$$

Note that for the $n = 2$ case, which is currently used in numerical simulations with $\alpha_0 = 1$, we get a residual volume-element ratio $\sqrt{\gamma_L/\gamma_0}$ of about

60%, meaning that the lapse vanishes when the gravitational collapse is still far from its final stage.

This is a first example of a limit surface. It occurs when and where the lapse collapse occurs before the metric collapse, that is,

$$\tau_L \;<\; \tau_S \,. \tag{6.59}$$

The final slice is then bounded away from the singularity, so that we can assume that both the space metric and the extrinsic curvature are regular tensors there. It follows that condition (6.54) can be replaced by

$$f\alpha\,|_{\alpha=0} \;<\; \infty\,, \tag{6.60}$$

implying that the limit surface will not be reached in a finite amount of coordinate time.

Hitting the limit surface, however, would not be a big problem, provided one avoids crossing the line and running into negative lapse values. Anyway, one must be very careful in order to avoid the divergence of some gauge-related quantities that could crash the numerical simulations:

- By rescaling some dynamical fields. The ordinary space derivative of the lapse
$$a_k \equiv \alpha\; A_k \tag{6.61}$$
could be used instead of the logarithmic one A_k. Also, the rescaled gauge factor
$$\tilde{f} = f\alpha^2 \tag{6.62}$$
could be used instead of the original one, and so on.
- By implementing some mechanism in the numerical code that avoids running into negative lapse values. The idea is to freeze the evolution of the set $\mathbf{u}$ of dynamical fields once the limit surface has been reached.

A completely different example of slicing condition leading to a limit surface is provided by the maximal slicing condition (2.96), that is,

$$trK = -\,\partial_\tau\; ln(\sqrt{\gamma}) = 0\,, \tag{6.63}$$

so that the space volume element is not even allowed to depart from its initial value. In the Schwarzschild case, maximal slicing is known to produce a limit surface given by [25]

$$r = r_L \equiv 3M/2 \;>\; 0\,, \tag{6.64}$$

where r is the Schwarzschild 'radial' coordinate, which is actually a time coordinate inside the horizon ($r = 2M$), so that (6.64) corresponds to a regular space-like hypersurface, bounded away from the $r = 0$ singularity.

We have seen that, when using maximal slicing, the lapse function must be computed by solving an elliptic differential equation (2.97), and this is why we have not considered it in the context of the hyperbolic evolution formalisms that appeared in the 1990s. Before these dates, maximal slicing was currently used in numerical simulations [26, 27] mainly because of its excellent singularity avoidance properties. It was actually the slicing of choice in the first 3D colliding black hole simulations [16].

6.2.3 Gauge pathologies

One can wonder whether singularity-avoidant slicing conditions can have some unwanted side effects. From the theoretical point of view, it is clear that the direct relationship (6.51) between (the proper time variations of) the lapse and the volume element will be helpful in the collapse scenario (6.45), where the lapse diminishes in response to the positivity of trK.

But, conversely, this will make the lapse function to increase in the zones which, for any reason, are expanding so that trK becomes negative. We will have then a sort of 'runaway' solution, with the time slicing accelerating precisely in the expanding regions. The lapse could grow then without limit, leading to a blowup.

One could wonder about how this blowup can arise in strongly hyperbolic systems, where the principal part can be expressed locally as a set of advection equations (4.68). The answer is precisely in the non-principal terms, which should be added in the right-hand side of the characteristic evolution equations, namely

$$\partial_t\, w_r + \lambda_r\, \partial_x\, w_r = q_r(w_s)\,, \tag{6.65}$$

where the characteristic source terms q_r are quadratic in the eigenfields w. The case of the gauge eigenfields (4.62) has been studied in 1D cases: gauge blowups can arise unless the condition [28, 29]

$$f = 1 + \frac{\text{const}}{\alpha^2} \tag{6.66}$$

is satisfied. Note that, allowing for the regularity requirement (6.60), condition (6.66) can only be satisfied in the harmonic slicing case ($f = 1$). This confirms that gauge blowups can actually occur in the generic singularity avoidance case, where $f > 1$.

As explained before, this behavior is generic in an expanding scenario ($trK < 0$), where a singularity avoidance condition is clearly a bad choice. But that blowup can arise even in collapse scenarios, where numerical errors can produce (spurious) negative values of trK. A remedy for this could be the introduction of some 'offset value' $K_0 > 0$ in the lapse evolution equation (6.51), for instance,

$$\partial_t\, ln\, \alpha = -f\alpha\,(tr\, K + K_0)\,, \tag{6.67}$$

so that the lapse will keep collapsing for $tr\,K > -K_0$. Of course, the offset value must be small enough to prevent the lapse collapse in the nearly static regions, where $trK \sim 0$.

6.3 Numerical Black Hole milestones

We will present in this section the results of applying the techniques discussed so far to black hole simulations. The idea is to choose a simple example and to follow it in a step-by-step way. We will therefore avoid unnecessary complications and go directly to the hearth of the problem, where much insight can be gained. The natural choice is that of a single Schwarzschild black hole, because it is the simpler one that shows the first difficulties that we will encounter in generic black hole simulations. We will also choose normal coordinates (zero shift). This apparently simple choice turns out to be a tough challenge from the numerical point of view, as we will see in what follows.

There are some decisions to take at the very beginning. These are our choices:

- **Evolution system**: The flux-conservative first-order version of the Z4 system, as detailed in Sect. 4.3.3.
- **Numerical algorithm**: The method of lines, with a third-order Runge–Kutta algorithm for the time discretization and the finite differences FDOC formula (5.93) for the space discretization.
- **Boundary conditions**: Maximally dissipative algebraic conditions at the outer boundaries, as described in Sect. 4.4.2.
- **Initial data**: Stuffed black hole initial data, with the scalar field given by (6.36).
- **Coordinate conditions**: Singularity-avoidant slicing, given by the algebraic condition (4.5), with zero shift. We keep all options open for the values of the gauge parameters f and m.

6.3.1 Short-term simulations

In keeping with the philosophy of this book, we will restrict ourselves to modest size simulations that can be performed in any personal computer. If we consider a single processor, which can usually address up to 2 Gb of memory, this implies using small numerical grids of, let us say, 80^3 nodes and being ready for overnight calculations in the worst case. Slightly larger grids can be used in today's multiple processor PCs, but then one must be able to do parallel computing. The Cactus computational toolkit [30] can provide some help by taking care of the required programming infrastructure.

Even if we are using cartesian 3D coordinates for a spherically symmetric problem, we can take advantage of some discrete space symmetries. The initial data are mirror symmetric across every coordinate plane. This reflection symmetry is preserved by the evolution equations, and this allows us to compute just over one octant of the spacetime domain. To illustrate how it can be done, let us focus for instance on the mirror symmetry across the z plane:

- We must classify every dynamical field as being either even or odd under the transformation
$$z \quad \longleftrightarrow \quad -z\,. \tag{6.68}$$
- We can then replace the negative z region of the numerical grid by a single layer (or two, depending on the stencil) of mirror points.
- The dynamical field values at these mirror points are defined to be the same ones as the corresponding ones in the corresponding layer of the positive z region: the sign of the odd fields is just reversed.

Octant symmetry allows us to double space resolution (that is, a factor 8 in memory space) for a given grid size. In our case, we will put the outer boundary at about $\rho = 10M$, which is definitely too close, in order to get a reasonable resolution, that is,

$$\Delta x = \Delta y = \Delta z = 0.1M\,. \tag{6.69}$$

Notice that a value of $10M$ along the coordinate axes means about $14.4M$ along the cube diagonals, which is the cut that we will show, unless otherwise stated, in the figures that follow. This will be enough to begin with.

6.3.1.1 Lapse collapse and landing (0–5M)

The first stages of the simulation show a collapse of the space volume element which, allowing for the singularity avoidance properties of the gauge conditions discussed in this chapter, translates itself into a collapse of the lapse function α. This lapse collapse can be slower or faster, depending on our choice of gauge parameters.

We show in Fig. 6.3 the lapse collapse for two choices of the gauge parameter f, with and without limit surface ($f = 2/\alpha$ and $f = 2$, respectively). We have taken the second gauge parameter $m = 2$ in both cases. The pattern is very similar: a fast collapse at the beginning (the lines shown are evenly spaced in time), followed by a slowing down and eventually 'landing' very close to a zero value in the innermost region, which means that the dynamics gets locally frozen there.

This is the first milestone in a black hole simulation. We have not shown the equivalent results for the harmonic slicing case because our code crashed before reaching that point. Notice that this was just the limit case at which the slicing still should manage to avoid the singularity at the continuum

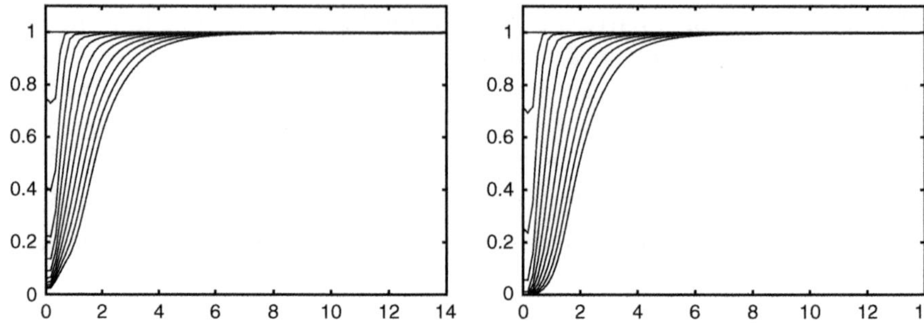

Fig. 6.3 Lapse collapse for different slicing conditions: the lapse values are shown every $0.5M$. The *left panel* corresponds to the $f = 2$ slicing, which has no limit surface. The *right panel* shows the same thing for the $f = 2/\alpha$ slicing, which does have a limit surface. The lapse collapses faster here, getting very close to zero in the innermost region, as it can be seen in the last plot ($t = 5M$).

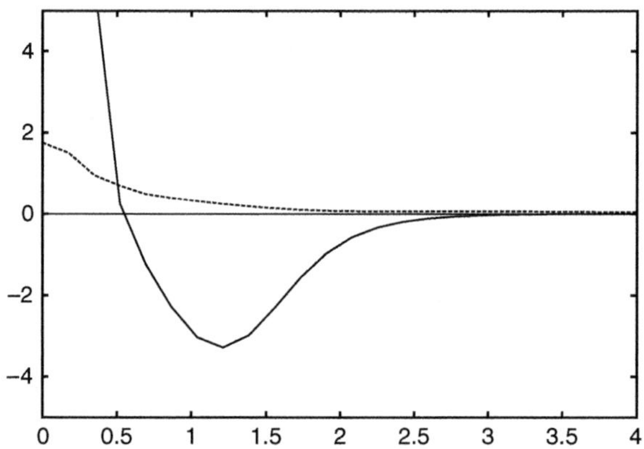

Fig. 6.4 Plots of K_{xx} (*solid line*) and trK (*dotted line*) along the x axis, corresponding to the $f = 2/\alpha$ slicing at $t = 6M$. Negative values of K_{xx} correspond to the increasing of γ_{xx}, that is, a radial stretching of the time slice. Notice that this is compatible with an overall collapse pattern, as shown by the positivity of trK: the radial stretching is then compensated by the collapse along the angular directions. Slice stretching is at the origin of an increasing lack of resolution which challenges the numerical algorithms.

level. In our case, numerical errors make the code to cross the line and the singularity is not avoided at the discrete level. Of course, different numerical algorithms could make numerical errors go into the opposite sense. But we will try to avoid here this kind of 'fine tuning', following instead more robust alternatives when available.

6.3.1.2 Slice stretching (4–20M)

Let us continue with the simulation, keeping an eye on the behavior of the extrinsic curvature, as shown in Fig. 6.4.

We see that trK is positive at that time (about $6M$ in our case), so that we have an overall collapse pattern. But if we look at a radial component (K_{xx} along the x direction for instance), we see that radial directions are actually expanding. The geometrical meaning of this behavior can be better understood by looking at the wormhole embedding diagram in Fig. 6.1. The radial expansion can be interpreted as the 'stretching' of the throat by pulling up from the top. This stretching is accompanied by the shrinking of the throat perimeter (the spherical surface area in the 3D case). The total effect on the space volume element is that of a highly inhomogeneous collapse. The radial expansion is at the origin of the lack of resolution that poses serious challenges to standard numerical algorithms.

We can see in Fig. 6.5 how a 'collapse front' is formed in, with a steep slope in the lapse profile. The same happens with trK, which can grow very quickly in the regions that are collapsing while keeping very low values in the neighboring ones. The resulting slope contrast can be more or less severe, depending on the selected slicing condition. When there is no limit surface, like in the $f = 2$ case, we can get short of resolution. Spurious oscillations can arise, even with our FDOC numerical method, leading to a code crash (remember that we can get very close to the singularity in those cases). The limit surface cases ($f = 1/\alpha$, $f = 2/\alpha$) are of course safer, because (small) numerical oscillations cannot drive us to the singularity. As we can see in the figure, the higher gauge-speed case ($f = 2/\alpha$) leads to a smoother profile, suitable for low-resolution simulations.

This is the second milestone we find in black hole simulations: getting rid of any spurious oscillations due to slice stretching. The way we have done it (at least in the limit surface cases), by using the FDOC algorithm, is by no means unique. In the $f = 2/\alpha$ case, the slopes contrast is not so dramatic and

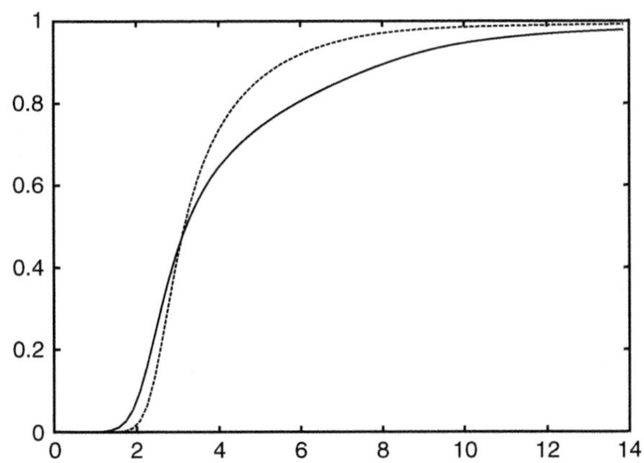

Fig. 6.5 Plots of the lapse profiles at $t = 10M$ for choices of the main gauge parameter, corresponding to different gauge speeds. The case $f = 2/\alpha$ (*solid line*) shows a smoother profile than $f = 1/\alpha$ (*dotted line*).

similar results are actually achieved in current black hole simulations just by adding higher order dissipative terms to a standard numerical algorithm [31]. Note, however, that slice stretching is a real dynamical feature that is still there, independent of the numerical algorithm one is using.

6.3.1.3 Lapse rebound (10–30M)

Let us see again what happens when allowing our simulations to proceed for a longer time (the precise value will depend both on the gauge and on the ordering parameter choices).

To monitor what is going on, let us take a look at Fig. 6.6, where we have plotted the same quantities as in Fig. 6.4. As a word of caution, let us remember that the lapse in the innermost region is already collapsed, so that the dynamics is frozen there. This means that the features we see in the innermost region in Fig. 6.6 correspond to an earlier stage (measured in proper time) than what we see around $\rho = 3M$, which is the region we are going to analyze now.

Contrary to what appeared in Fig. 6.4, we see here that trK is negative, so we have an overall expansion pattern of the space volume element, namely

$$(\partial_t - \mathcal{L}_\beta)\, ln(\sqrt{\gamma}) = -\alpha\, trK \; > \; 0\,. \tag{6.70}$$

At the continuum level ($\Theta = 0$), this would cause an increasing of the lapse, which is driven by the equation

$$(\partial_t - \mathcal{L}_\beta)\, ln(\alpha\,) = -\alpha\, f\,(trK - m\Theta)\,, \tag{6.71}$$

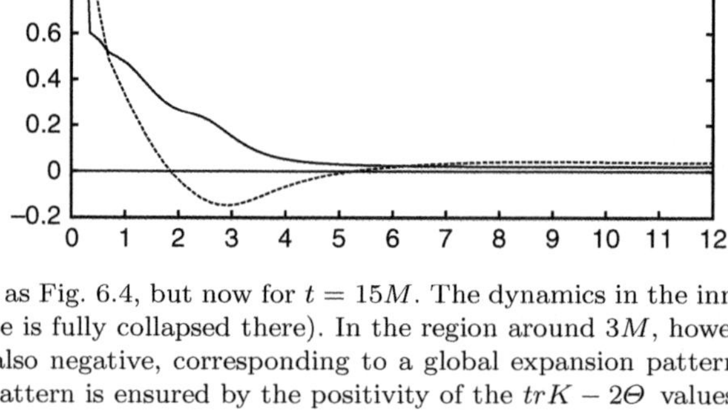

Fig. 6.6 Same as Fig. 6.4, but now for $t = 15M$. The dynamics in the innermost region is frozen (the lapse is fully collapsed there). In the region around $3M$, however, trK (*dotted line*) becomes also negative, corresponding to a global expansion pattern. Note that the lapse collapse pattern is ensured by the positivity of the $trK - 2\Theta$ values (*solid line*).

which could grow out of control, leading to a runaway solution. The singularity avoidance of the slicing condition (6.71) works against us in an expanding scenario, where $trK > 0$, as discussed in Sect. 6.2.3.

One can wonder why this rebound problem is not an issue in standard BSSN simulations. In order to explain this, remember that the BSSN evolution equation for trK corresponds to the Z4 evolution equation for the combination $trK - 2\Theta$ (this is actually why we took the second gauge parameter $m = 2$ in our simulations). This means that we may circumvent the rebound problem by making use of the energy–constraint violations in the black hole interior, which contribute to non-trivial Θ values. In our case, with the field equations solved consistently everywhere, this is done automatically by choosing $m = 2$.

This should work fine both for the stuffing and the punctures approaches. Note, however, that this will not be the case in the free black hole approach. Suppressing artificially the matter sources in the field equations amounts to adding some 'negative energy' counterterms that can reverse the sign of Θ there. Negative values of m may be required in this case.

6.3.1.4 Boundary conditions (30M and beyond)

Let us see once more what happens when allowing our simulations to proceed for an even longer time. We see in Fig. 6.7 that the lapse rebound does not show up. What we see is the collapse of the whole computational region, which ends up completely inside the apparent horizon. At this point, one must remember that the lapse function provides the scale factor between

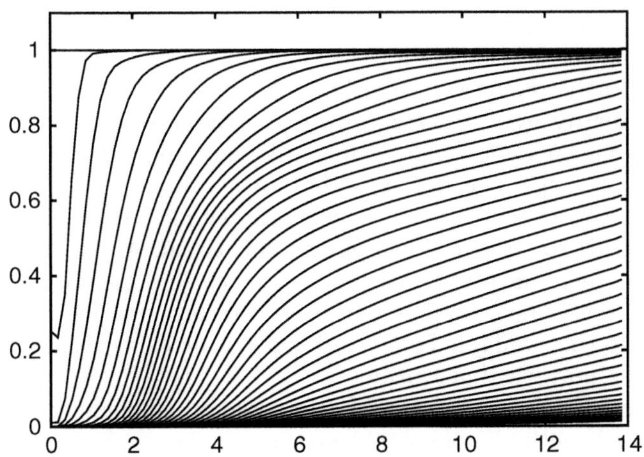

Fig. 6.7 Same as Fig. 6.5, but now running up to $50M$ (coordinate time at the outer boundary). The plots are shown now with a larger time interval in order to get a clearer picture. The whole computational region gets inside the black hole horizon.

coordinate and proper time: the lapse collapse in the outermost region in Fig. 6.7 means that proper time gets frozen there slightly above $40M$, whereas coordinate time goes till $50M$ and beyond.

When one reaches this point, two opposite thoughts come to mind:

- A positive one: the proposed calculation has been completed. The main obstacles (spurious oscillations, gauge problems) have been passed. The boundary conditions (algebraic ones, remember) are doing a good job, even if the outer boundary is placed definitely too close, at $\rho = 10M$. So, it is an achievement.
- A negative one: even if one could evolve the black hole simulation further, this would be of no value for extracting any physical consequence for the outside geometry. We have reached the limit of our (modest) computational resources.

6.3.2 Long-term simulations

As scientists, however, we will always try to overcome our limits. There is a number of ways in which we could do that in this case:

- **Getting more computational resources.** We can switch from our modest PC to a bigger computer, maybe just a PC cluster. This will allow to put the outer boundary farther away, so that we can model even black hole collisions and predict the waveforms of the resulting gravitational radiation. This is a limited improvement, however. Every factor 2 in resolution means a factor 2^3 in the number of nodes in a 3D grid (this means eight times the original memory requirements). Moreover, if one is using explicit finite differences methods, there is an extra doubling on the number of required time steps, allowing for the Courant condition. This amounts to a 16 factor in the required computing power.
- **Implementing even better numerical methods.** In the CFD arsenal, we can find for instance multi-patch and adaptive mesh refinement (AMR) methods [32, 33]. Instead of a homogeneous numerical grid, multi-patch methods make use of different coordinate patches (each one with its own adapted grid) to cover the computational region. AMR methods go even farther, by dynamically increasing (or decreasing) the grid resolution exactly where it is needed. These advanced techniques are devised in order to optimize the available computational resources at the cost of an extra load on the numerical algorithm.
- **Distorting the space slices.** Instead of adapting the numerical grid, like in the AMR methods, one could just adapt the space coordinate system (and keep a simple, homogenous grid). This can be done by a simple coordinate transformation. The coordinate position of our boundary can even

correspond to space infinity [34]. This 'fisheye transformation' idea [35, 36] has been actually implemented in black hole simulations.

- **Taking advantage of the shift degrees of freedom**. This means providing a dynamical shift choice that drives grid points to the places where more resolution is needed [37–39]. Also, in black hole simulations, one expects the spacetime going through a highly dynamical transient phase, then approaching a quasi-stationary final stage. Coordinate choices in which the time lines tend to follow the corresponding 'quasi-Killing' vector would be of a great help in these cases [40].

Let us consider here a combination of the first and third options mentioned above. Appealing to the space coordinates freedom, we will switch to some logarithmic coordinates, as defined by

$$\rho = L \, \sinh(R/L) \,, \tag{6.72}$$

where R is the new radial coordinate and L some length scale factor. We will perform a 200^3 points simulation: this will require 32 nodes of a cluster working for days just for us. We have performed a long-term numerical simulation for the $f = 2/\alpha$ case, with $L = 1.5M$, so that our boundary, placed at $R = 20M$ in these logarithmic coordinates, corresponds to about $\rho = 463.000M$ in the original isotropic coordinates. In this way, as shown in Fig. 6.8, the collapse front is safely away from the boundary, even at very late times. We stopped our code at $t = 1000M$, without any sign of instability.

These results provide a new benchmark for numerical relativity codes: a long-term simulation of a single black hole, without excision, in normal coordinates (zero shift). Moreover, it shows that a non-trivial shift prescription is not a requisite for code stability in black hole simulations. From the numerical point of view, it is an excellent example of what can be done with a single numerical grid.

6.3.3 Further developments

Beyond the basic developments presented in this chapter, in which we have consider a single black hole evolution, there is the obvious next step: a binary black hole (BBH) system. The study of the late orbital stage of BBH systems is of particular interest, because they are among the most likely sources to be detected by the current gravitational wave observatories [41] and by the space-based project LISA [42].

Initial data for BBH systems in quasi-circular orbits can be obtained in many ways [1]:

- In the punctures approach, one can take advantage of the linearity of (6.19) by just adding different punctures (Brill–Lindquist data) [43, 44]. The momentum and spin of every puncture can be incorporated by means of a

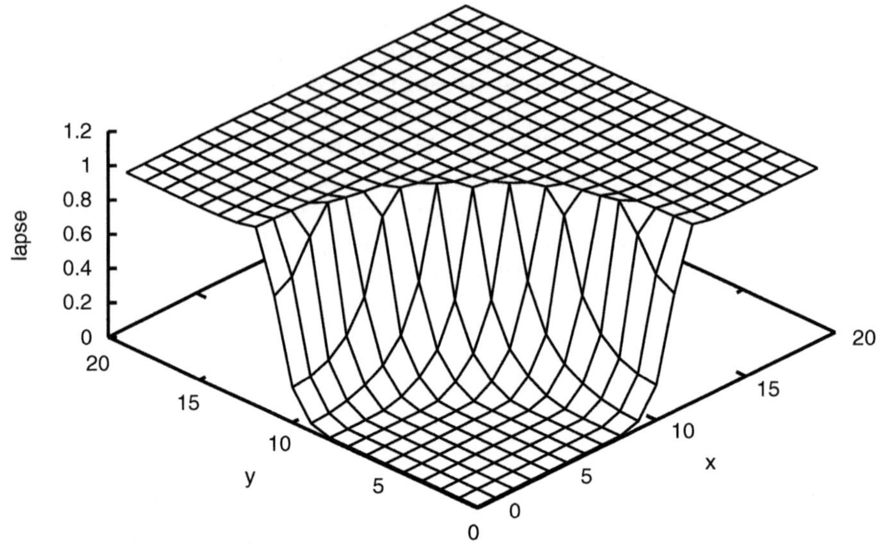

Fig. 6.8 Plot of the lapse function for a single BH at $t = 1000M$ in normal coordinates. Only 1 of every 10 points is shown along each direction. The third-order accurate FDOC algorithm has been used with $b = 1/12$ and a space resolution $h = 0.1M$. The profile is steep, but smooth: no sign of instability appears. *Small riddles*, barely visible on the *top* of the collapse front, signal some lack of resolution because of the logarithmic character of the grid. The dynamical zone is safely away from the boundaries.

non-vanishing extrinsic curvature [45]. Finally, the data can be compactified in the interior asymptotically flat region [6].

- In the excision approach, one can take advantage of the invariance of the Kerr–Schild data under boost transformations [46]. This allows to use them in a (boosted) superposed form as a initial guess for a constraint-solving process. The amount of constraint violations is usually so small that even the unconstrained data can be used as a starting point for numerical simulations [47].
- With the goal of getting astrophysically realistic initial data, the second post-Newtonian approximation has been considered as a starting point [48, 49]. The resulting initial data can be shown to agree with the exact solutions of the constraint equations, modulo a coordinate transformation, up to the required post-Newtonian order [50].

BBH simulations are very demanding from the computational point of view. On the one hand, one must resolve the steep gradients near the individual BH horizons, typically requiring resolutions of (at least) $0.1M$. On the other hand, the calculation of accurate waveforms requires to extract waves at sufficient large radii, ideally in the wave zone. This implies using computational grids at least two orders of magnitude larger than the radius of a single BH. With the current computational power, this can only be achieved by using some form of mesh refinement, including the implementation of some kind of 'fisheye' space coordinates, as discussed in the previous section.

The first major breakthrough in BBH numerical simulations was reported in [51]. Using a careful choice of co-rotating shift and excision, they where able to evolve a BBH for a full quasi-circular orbit. However, it was not possible to extract the merger waveform. The first fully numerical evaluation of the waveform from a BBH was reported by Pretorius [7]. By using the generalized harmonic framework, combined with special numerical techniques, he evolved a system in which two scalar fields collapsed to form individual BH's, which then formed a merging elliptical binary. These results were immediately followed by analogous developments in the BSSN framework [8, 9], leading to the 'moving punctures' approach.

The accuracy of this approach allows to address some relevant astrophysical issues, like the radiation of linear and angular momentum by unequal mass systems, including the gravitational radiation recoil (kick) on the remnant object [52, 53]. In parallel, the joint effort toward gravitational wave detection continues. Current BBH signal templates come from the post-Newtonian approximation, focusing on the inspiral phase. More recently, hybrid templates are being produced, incorporating numerical relativity results for the merger phase. This allows a cross-validation between analytical and numerical results in the transition regime from the inspiral to the merger phase (see for instance [54] and references therein). This can be done now for many mass ratios and spin configurations. And old bet has been won by numerical relativists: our templates are ready, waiting for signal detection by experimentalists.

References

1. G. B. Cook, Initial data for numerical relativity, Liv. Rev. Relat. http://www.livingreviews.org/lrr-2000-5
2. J. Thornburg, Event and apparent horizon finders for 3 + 1 numerical relativity. Liv. Rev.Relat. http://www.livingreviews.org/lrr-2007–3
3. A. Lichnerowicz, J. Math. Pures et Appl. **23**, 37 (1944).
4. R. Beig and N. O Murchadha, Class. Quantum Grav. **11**, 419 (1994).
5. R. Beig and N. O Murchadha, Class. Quantum Grav. **13**, 739 (1996).
6. S. Brandt and B. Brügmann, Phys. Rev. Lett. **78**, 3606 (1997).
7. F. Pretorius, Phys. Rev. Lett. **95**, 121101 (2005).
8. M. Campanelli, C. O. Lousto, P. Marronetti and Y. Zlochower, Phys. Rev. Lett. **96**, 111101 (2006).
9. J. G. Baker et al., Phys. Rev. Lett. **96**, 111102 (2006).
10. P. Anninos et al., Phys. Rev. Lett. **74**, 630 (1995).
11. J. Libson et al., Phys. Rev. **D53**, 4335 (1996).
12. A. Arbona et al., Phys. Rev. **D57**, 2397 (1998).
13. D. Alic, C. Bona and C. Bona-Casas, Phys. Rev. **D** (2009). ArXiv:0811.1691
14. D. Brown et al., Phys. Rev. **D76**, 081503(R) (2007).
15. J. Thornburg, Class. Quan. Grav. **4**, (1987).
16. P. Anninos et al., Phys. Rev. Lett. **71**, 2851 (1993).

17. C. Bona, J. Massó, E. Seidel and J. Stela, Phys. Rev. Lett. **75**, 600 (1995).
18. C. Bona, J. Massó, E. Seidel and J. Stela, Phys. Rev. **D56**, 3405 (1997).
19. E. Seidel and W. M. Suen, Phys. Rev. Lett. **69**, 1845 (1992).
20. B. Szilágyi, Class. Quantum Grav. **24**, S275–S293 (2007).
21. H. P. Pfeiffer, Class. Quantum Grav. **24**, S59–S81 (2007).
22. F. Pretorius, Class. Quantum Grav. **23**, S529 (2006).
23. R. A. Matzner, M. F. Huq and D. Shoemaker, Phys. Rev. **D59**, 024015 (1998).
24. C. Bona and J. Massó, Phys. Rev. **D38**, 2419 (1988).
25. F. Estabrook et al., Phys. Rev. **D7**, 2814 (1973).
26. L. Smarr and J. W. York, Phys. Rev. **D17**, 1945 (1978).
27. L. Smarr and J. W. York, Phys. Rev. **D17**, 2529 (1978).
28. B. Reimann, M. Alcubierre, J. A. González and D. Núñez, Phys. Rev. **D71**, 064021 (2005).
29. M. Alcubierre et al., Phys. Rev. **D72**, 124018 (2005).
30. http://www.cactuscode.org
31. B. Gustafson, H. O. Kreiss and J. Oliger, *Time Dependent Problems and Difference Methods*, Wiley, New York (1995).
32. M. J. Berger and R. J. LeVeque, SIAM J. Numer. Anal. **35**, 2298 (1998).
33. F. Pretorius and L. Lehner, J. Comput. Phys. **198**, 10 (2004).
34. H. Friedrich, 'Conformal Einstein evolution' in *The Conformal Structure of Space-Times: Geometry, Analysis, Numerics*, ed. by J. Frauendiener and H. Friedrich, Springer Lecture Notes in Physics, Vol 604, pp. 1–50, Springer, Berlin Heidelberg New York (2002).
35. J. Baker, B. Brügmann, M. Campanelli and C. O. Lousto, Class. Quantum Grav. **17**, L149 (2000).
36. J. Baker, M. Campanelli and C. O. Lousto, Phys. Rev. **D65**, 044001 (2002).
37. M. Alcubierre and B. Brügmann, Phys. Rev. **D63**, 104006 (2001).
38. M. Alcubierre et al., Phys. Rev. **D67**, 084023 (2003).
39. L. Lindblom and M. A. Scheel, Phys. Rev. **D67**, 124005 (2003).
40. C. Bona, J. Carot and C. Palenzuela-Luque, Phys. Rev. **D72**, 124010 (2005).
41. http://www.ligo.caltech.edu
http://www.virgo.infn.it
http://geo600.aei.mpg.de
http://tamago.mtk.nao.ac.jp
http://www.gravity.uwa.edu.au
42. K. Dantzmann and A. Rudiger, Class. Quantum Grav. **20**, S1 (2003).
43. C. Misner and J. Wheeler, Ann. Phys. (N.Y.) **2**, 525 (1957).
44. D. R. Brill and R. W. Lindquist, Phys. Rev. **131**, 471 (1963).
45. J. M. Bowen and J. W. York, Phys. Rev. **D21**, 2047 (1980).
46. E. Bonning, P. Marronetti, D. Neilsen and R. A. Matzner, Phys. Rev. **D68**, 044019 (2003).
47. P. Marronetti et al., Phys. Rev. **D62**, 024017 (2000).
48. L. Blanchet, Phys. Rev. **D68**, 084002 (2003).
49. W. Tichy, B. Brügmann, M. Campanelli and P. Diener, Phys. Rev. **D67**, 064008 (2003).
50. S. Nissanke, Phys. Rev. **D73**, 124002 (2006).
51. B. Brügmann, W. Tichy and N. Jansen, Phys. Rev. Lett. **92**, 211101 (2004).
52. F. Herrmann, I. Hinder, D. Shoemaker and P. Laguna, Class. Quantum Grav. **24**, S33–S42 (2007).
53. J. G. Baker et al., Astrophys. J. Lett. **653**, 93–96 (2006).
54. A. Gopakumar, M. Hannam, S. Husa and B. Brügmann, Phys. Rev. D78, 064026 (2008).

Chapter 7
Matter Spacetimes

Up to now we have been considering mainly vacuum spacetimes. Although this has been useful to understand the dynamics of Einstein equations, most of the realistic scenarios involve some kind of matter. For instance, at large scales we have cosmological models, based on the isotropic distribution of dust (i.e., non-interacting particles). At intermediate scales there are galaxy models for dark matter, which has been modeled by using either dust or scalar fields. A small-scale approach would include all types of astrophysical compact objects and the dynamics related to them; binary star evolution, core collapse, accretion disks, etc.

Einstein's theory describes how the spacetime is deformed by the matter and how the matter moves over this curved spacetime. The evolution of the spacetime is given by the Einstein field equations and, allowing for the Bianchi identities, this implies that the evolution of matter must comply with the conservation of the stress–energy tensor

$$\nabla_\nu T^{\mu\nu} = 0. \tag{7.1}$$

In addition to these equations, which only ensure the conservation of energy and momentum, the matter evolution may be subject to other physical restrictions. For instance, if the matter is made of baryons, and if it is neither created nor destroyed, then it must conserve also baryon density. Another example comes from electromagnetic fields, since Maxwell equations cannot be fully derived from the four conditions (7.1) and must be imposed separately. We will address properly these points in this chapter.

We are going to consider some standard stress–energy tensors describing different types of matter. Some of these matter models allow stationary stable solutions which can describe several astrophysical scenarios. We will start by the simplest one, composed of scalar fields, which can be used as a model for dark matter in galaxies. We will follow with the electromagnetic fields, where simple models can be considered as a preliminary step for further developments. Later we will consider perfect fluids, which reproduce the expected

Bona, C. et al.: *Matter Spacetimes*. Lect. Notes Phys. **783**, 171–209 (2009)
DOI 10.1007/978-3-642-01164-1_7

behavior of astrophysical plasmas in most types of stars. Finally, the Maxwell equations are coupled to the fluid by means of the MHD approximation, providing a general framework for studying magnetized (neutron) stars.

7.1 Scalar fields

Probably one of the simplest fluids one can imagine is the one composed by a scalar field. For later convenience we will consider a complex scalar field Φ, which obeys a generalized wave equation in a curved background, under the effect of some potential $V(|\Phi|)$. This system is commonly known as the (generalized) Klein–Gordon equation, and it allows for stable configurations when the dispersive character of the wave equation is balanced by the gravitational attraction. These regular solutions are called boson stars and describe a family of self-gravitating scalar field configurations within general relativity. One can in this way consider strongly gravitating compact star spacetimes without the worries associated with weak solutions, such as shock fronts and discontinuities in the fluid variables. This makes boson stars a very useful probe of strong-field general relativity.

7.1.1 The Klein–Gordon equation

The dynamics of a complex scalar field in a curved spacetime is described by the following stress–energy tensor

$$T_{\mu\nu} = \frac{1}{2}\left[\nabla_\mu\overline{\Phi}\,\nabla_\nu\Phi + \nabla_\mu\Phi\,\nabla_\nu\overline{\Phi}\right] - \frac{1}{2}g_{\mu\nu}\left[\nabla_\lambda\overline{\Phi}\,\nabla^\lambda\Phi + V\left(|\Phi|^2\right)\right] \quad (7.2)$$

where Φ is the scalar field, $\overline{\Phi}$ its complex conjugate, and $V(|\Phi|^2)$ a potential depending only on $|\Phi|^2$. When the scalar field is real then $\Phi = \overline{\Phi}$ and this stress–energy tensor can be written in a simpler way, as we did in the previous chapter. Note also that for scalar quantities $\nabla_\mu\Phi = \partial_\mu\Phi$.

In the complex case, one can take advantage of the conserved current

$$J^\mu = \frac{i}{2}\,g^{\mu\nu}\left[\overline{\Phi}\,\nabla_\nu\Phi - \Phi\,\nabla_\nu\overline{\Phi}\right], \quad (7.3)$$

which verifies the divergence equation

$$\nabla_\mu J^\mu = \frac{1}{\sqrt{g}}\,\partial_\mu[\sqrt{g}J^\mu] = 0\,, \quad (7.4)$$

which ensures the conservation of the density $N \equiv J^\mu n_\mu$. The corresponding charge (i.e., the space volume integral of N) can be interpreted as the boson particle number [1]. These results can also by obtained by applying the Noether theorem to the invariance of the stress–energy tensor under phase shifts of the scalar field: N would then be the associated Noether charge density.

The evolution equations for the scalar field can be obtained by taking the divergence of this stress–energy tensor. The resulting system is known as the (generalized) Klein–Gordon equations, which can be written as

$$\Box \Phi = \frac{dV}{d|\Phi|^2}\, \Phi\,. \tag{7.5}$$

From now on we will consider for simplicity the free field case, where the potential takes the simple form

$$V(\,|\Phi\,|^2\,) = m^2\ |\,\Phi\,|^2\,, \tag{7.6}$$

with m a parameter that can be identified with the bare mass of the field theory, although it has units of inverse length. Dimensionless units would amount to fix this mass value to unity.

For a numerical implementation, it is also useful to reduce (7.5) to a fully first-order system, along the lines discussed in Chap. 5. The reduction in time is achieved by introducing new independent variables related to the time derivatives of the fields. We define

$$\Pi \equiv n^\mu\ \partial_\mu \Phi\,, \tag{7.7}$$

where we are using here again the unit normal n_μ to the time slices. The evolution of Π is now given by (the first-order form of) the KG equations (7.5), while the evolution of Φ is simply given in terms of Π by the definition (7.7).

The reduction to first order in space is made as usual, by introducing new independent variables encoding the first space derivatives as

$$D_i \equiv \partial_i \Phi. \tag{7.8}$$

The equations of motion for this first-order quantity can be obtained along the lines discussed in Chap. 4. Notice that one encounters in this way 'first-order constraints', namely

$$\mathcal{C}_i \equiv D_i - \partial_i \Phi = 0\ , \tag{7.9}$$

which must be satisfied for a consistent solution. At this point the resulting first-order system is described by the evolution equations for the array of fields $\{\Phi, \Pi, D_i\}$, together with the first-order constraints (7.9).

We can take advantage here of the results we presented for the generalized harmonic system in Sect. 4.3.2. The full first-order reduction of the Klein–Gordon equations can be written then as

$$
\begin{aligned}
\partial_t \Phi &= \beta^k D_k - \alpha\, \Pi && (7.10)\\
\partial_t \Pi - \beta^k \partial_k \Pi &= -\alpha\, \gamma^{ij} \partial_i D_j - \kappa\, \beta^k\, (\partial_k \Phi - D_k) && (7.11)\\
&\quad - \frac{\alpha}{2} \Pi (n^\mu n^\nu Q_{\mu\nu}) + \alpha \gamma^{ij}\, D_i (n^\mu Q_{j\,\mu}) + \alpha\, \Gamma_\mu\, \Phi^\mu + \alpha\, m^2\, \Phi \\
\partial_t D_i - \beta^k \partial_k D_i &= -\alpha\; \partial_i \Pi + \alpha\; \kappa\, (\partial_i \Phi - D_i) && (7.12)\\
&\quad + \frac{\alpha}{2}\; \Pi (n^\mu n^\nu D_{i\mu\nu}) - \alpha\; \gamma^{jk} D_k (n^\mu D_{ij\mu}),
\end{aligned}
$$

where κ is the damping parameter for the first-order constraints (7.9). Equations (7.10), (7.11), and (7.12) constitute the Klein–Gordon (KG) system in our implementation.

The characteristic decomposition, along any given space direction n^k, for the KG system (7.10), (7.11), and (7.12) is given by the following eigenmodes:

$$
\begin{aligned}
&\Phi && (\text{speed} \;\; 0) \\
D_\perp &\equiv D_i - D_n\; n_i && (\text{speed} \;\; -\beta^n) && (7.13)\\
w^{(\pm)} &\equiv \Pi - \kappa\; \Phi \pm D_n && (\text{speed} \;\; -\beta^n \pm \alpha),
\end{aligned}
$$

where again the index n indicates contraction with the (unit) vector n^k. Thus the incoming modes at the outer boundary are given by w^- and $D_\perp$, where $\beta^n > 0$.

7.1.2 Boson stars initial data

As discussed before, the boson star is a stationary solution of the Einstein–Klein–Gordon equations, where the dispersive character of the scalar field is balanced with the gravitational attraction. We will restrict ourselves to the simplest case, which is obtained by assuming spherical symmetry, so the initial data for the boson star can be computed with a 1D code. The resulting solution can be easily written then in 3D cartesian-like coordinates. This 1D solution can be obtained from the following ansatz for the scalar field:

$$
\Phi(t, r) = \phi\,(r)\; e^{-i\omega t}. \tag{7.14}
$$

With this assumption, our goal is then to find the amplitude $\phi\,(r)$, the frequency ω, and the metric coefficients, such that the spacetime generated by this matter configuration is static.

The procedure gets a simpler formulation in Schwarzschild coordinates [2, 3]. The line element then takes the form

$$
ds^2 = -\alpha\,(r)^2\, dt^2 + a\,(r)^2\, dr^2 + r^2 d\Omega^2. \tag{7.15}
$$

The equilibrium conditions in this coordinate system are then given by

$$
\begin{aligned}
a' &= \frac{a}{2}\left\{-\frac{a^2-1}{r} + 4\pi r\left[\left(\frac{\omega^2}{\alpha^2}+m^2\right)a^2\phi^2 + \phi'^2\right]\right\}, \\
\alpha' &= \frac{\alpha}{2}\left\{\frac{a^2-1}{r} + 4\pi r\left[\left(\frac{\omega^2}{\alpha^2}-m^2\right)a^2\phi^2 + \phi'^2\right]\right\}, \qquad (7.16) \\
\phi'' &= -\left(1+a^2-4\pi r^2 a^2 m^2 \phi^2\right)\frac{\phi'}{r} - \left(\frac{\omega^2}{\alpha^2}-m^2\right)\phi\, a^2,
\end{aligned}
$$

where the primes denote differentiation with respect to r. In order to obtain a physically consistent solution, we will impose the following boundary conditions, which guarantee both regularity at the origin and asymptotic flatness:

$$a\,(0) = 1, \quad \phi'\,(0) = 0, \qquad \alpha'\,(0) = 0, \qquad (7.17)$$

$$\lim_{r\to\infty} a\,(r) = 1, \quad \lim_{r\to\infty} \phi\,(r) = 0, \quad \lim_{r\to\infty} \alpha\,(r) = 1\,. \qquad (7.18)$$

Note that these asymptotic conditions replace the corresponding ones for ordinary stars, where the star radius is defined by the vanishing of pressure. In boson stars one cannot expect such pressure vanishing, since the scalar field decays exponentially up to infinity: the star radius is rather defined by the surface containing a significative percentage of the mass–energy, let us say, 95%. We are using here the mass function, which can be defined for any spherically symmetric spacetime as

$$2M = Y\left[1 - g^{\mu\nu}\partial_\mu Y \partial_\nu Y\,\right], \qquad (7.19)$$

where Y is the area radius. In Schwarzschild coordinates, $Y = r$, we have

$$M(r) = \frac{r}{2}\,(1 - 1/a^2)\,. \qquad (7.20)$$

This function gives the value of the total mass enclosed in a sphere of radius Y in the given spacetime [4].

For a given value of the central amplitude $\phi\,(0) = \phi_c$ those equilibrium equations and boundary conditions only admit solutions for a discrete set of ω values. This is a sort of eigenvalue problem, where both the eigenfunction and its eigenvalue must be determined at the same time. In our particular case, we are interested on the fundamental (lowest frequency) solution, where the scalar field profile contains no nodes.

The problem is solved by integrating from $r = 0$ outward using a second-order shooting method (see for instance [5]), aiming for a monotonically decreasing $\phi(r)$ lower than some threshold at $r = r_{\mathrm{max}}$. The system is integrated for an arbitrary initial value of the lapse ($\alpha(0) = 1$ for instance) and with an initial guess for ω, and the iterations proceed until the value $\phi(r_{\mathrm{max}})$ gets below the given threshold. After all the equations are solved, the lapse is rescaled in order to obtain a function which asymptotes to 1 at infinity.[1] The same rescaling must of course be performed with the frequency ω.

Once the solution is computed in this coordinate system a change of coordinates is performed to isotropic ones:

$$ds^2 = -\alpha^2(\rho)\, dt^2 + \psi^4(\rho)\left(d\rho^2 + \rho^2 d\Omega^2\right). \tag{7.21}$$

In these coordinates the extension to three dimensions is direct since the space part is explicitly conformally flat, that is,

$$d\rho^2 + \rho^2 d\Omega^2 \qquad \leftrightarrow \qquad \delta_{ij}\, dx^i dx^j\,. \tag{7.22}$$

The values of the radial functions in the cartesian grid can be computed numerically, as in [6]. We can obtain in this way initial data for $g_{\mu\nu}$ and ϕ, the rest of the fields for the 3D code are chosen as follows: $Q_{\mu\nu} = 0$, Π from the ansatz (7.14), and the space derivatives $D_{i\mu\nu}$ and D_i from the first-order constraints (7.8). This completely defines the initial data for a boson star.

The potential (7.6) leads to the so-called *miniboson stars*, because achievable (stable) configurations have small masses. More general terms can be included, such as the $\lambda\ |\phi|^4$ self-interaction term introduced in [7], leading to heavier *boson stars* which have masses and sizes more relevant to astrophysical applications. The mass diagrams are plotted in Fig. 7.1. On the left panel, the mass is shown as a function of the scalar field at the origin ϕ_c. The stable branch is the region located on the left side of the curve, satisfying

$$\frac{\partial M}{\partial \phi_c} \geq 0. \tag{7.23}$$

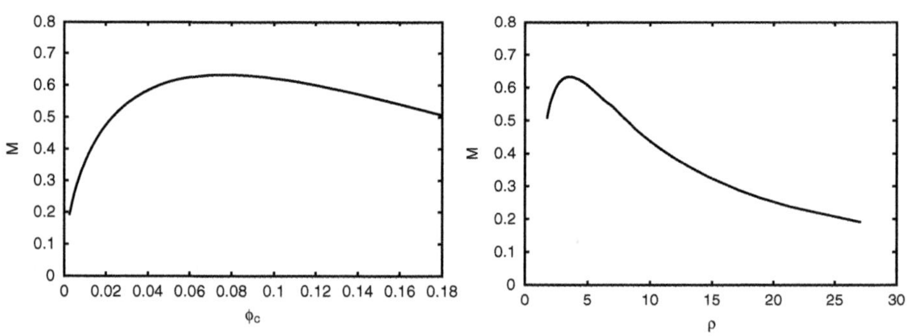

Fig. 7.1 The mass of the boson star as a function of the central value of the scalar field (*left panel*) and of the radius (*right panel*). The maximum stable mass is at $M = 0.633$. The stable branch is at the *left (resp. right) side* of the maximum in the *left (resp. right) panel*

[1] This can be done because of the linearity of the lapse equation (7.16), assuming that ω scales with α.

Boson star configurations on this branch are stable against small perturbations. On the right panel in Fig. 7.1 the mass is shown instead as a function of the radius. On the stable branch (here at the right of the maximum), we can see that heavier boson stars have a smaller radius since the gravitational attraction is larger, compressing the scalar field, being balanced only by its dispersive character and the quadratic potential. Notice that the difference on the radius between two stars with a factor 2–3 in the mass can be of one order of magnitude or even more.

Boson star configurations in the opposite branch are unstable against small perturbations. The final state will depend on the binding energy of the system, which is defined as

$$E_{\text{binding}} = M - Nm, \tag{7.24}$$

where N is the number of bosons given by the Noether charge. If the binding energy is positive, the boson star will disperse some scalar field and may decay to a configuration on the stable branch, while it will collapse to a black hole in the other case.

The profile of the boson star used for evolution in the next section is shown here as an example. The scalar field, the lapse, and the metric components are shown in Fig. 7.2.

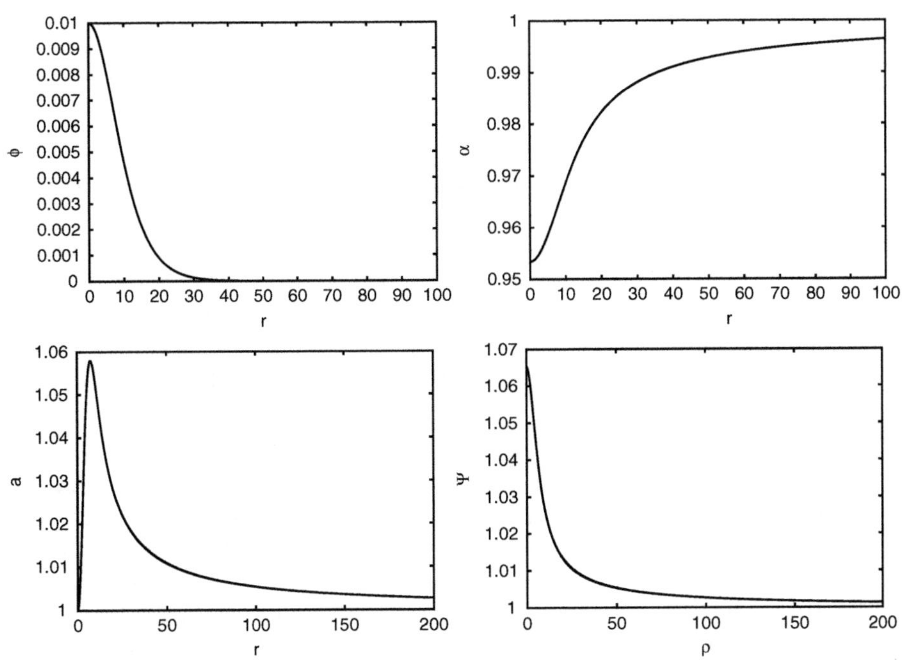

Fig. 7.2 Profiles of the scalar field, the lapse, and the metric coefficients (in Schwarzschild and isotropic coordinates) for a boson star in the stable branch.

7.1.3 Evolution of a single boson star

We will perform here single boson star evolution simulations, as a first step to the study of more complex systems. We begin with initial data describing a star with the central value of the scalar field amplitude being $\phi_c = 0.01$. This star has total mass $M_{1D} = 0.361$ and radius $R_{95} = 13$, well inside the stable branch of solutions. The corresponding frequency is $\omega_{1D} = 0.976$.

We will use the generalized harmonic system for the time evolution. Note at this point that we are trying to model a static equilibrium solution. Our time coordinate can be adapted to the associated Killing vector, getting a harmonic slicing as a result. This can be checked easily in normal coordinates, where the harmonicity of the time coordinate is given by

$$\Box x^0 \qquad \leftrightarrow \qquad \partial_t(\alpha/\sqrt{\gamma}) = 0\,. \tag{7.25}$$

Concerning the space coordinates, however, there is a conflict between the harmonicity condition and the staticity one, which would amount in this case to consider normal coordinates, orthogonal to the time Killing vector. Put in another way, the gauge sources corresponding to our initial data verify (normal coordinates)

$$H^0 = 0\,, \qquad H_i = H_i(0, x) \neq 0\,. \tag{7.26}$$

In view of this conflict, we will consider two options for performing the evolution:

- Keeping the staticity requirement. Static gauge source functions $H^a(x)$ must be added, keeping the initial values (7.26), which will then hold for all times.
- Enforcing strictly harmonic conditions ($H^\mu = 0$). In this case there will be a fictitious evolution of the system due to the choice of non-adapted coordinates. This will be useful in order to test the ability of the harmonic coordinate condition for adapting itself to the physical problem under consideration. This will be important for more complicated non-static problems, like the binary boson star case.

The evolution of an isolated star is performed for these two different coordinate systems. The values of a scalar quantity should agree at the center of symmetry for both the harmonic and the normal (static) coordinates. This implies that the scalar field should have the same local oscillatory behavior (7.14) at $r = 0$ in both coordinate systems. This is indeed the case and is illustrated in Fig. 7.3, where we plot the value at the origin of (the real part of) the scalar field, displaying the expected oscillatory behavior with a frequency $\omega = 0.96 \pm 0.03$. This is in very good agreement with the frequency obtained from the 1D initial data.

Some coordinate effects do arise, however. As discussed above, we get some non-trivial coordinate-induced dynamics since we are not adopting a coordinate system in which the spacetime is explicitly static. This effect can be seen in Fig. 7.4 where the evolution of the maximum of g_{xx} as a function of coordinate time t for three different resolutions is displayed. As is evident in the figure, there is an initial transient variation of the metric coefficient, which later approaches a constant value.

7.2 Electromagnetic fields

A number of fascinating phenomena are mediated by electromagnetic fields and their interaction with other physical processes in a given gravitational field source. This interaction may significantly affect the dynamics of the source, which can produce radiation that could be detected far away.

The evolution of the magnetic fields is given by the Maxwell equations, which can be written in terms of either the vector potential or the electromagnetic fields themselves. We will review in this section both alternatives, paying special attention to the treatment of the constraints and the hyperbolicity of the equations. Notice that, in the context of general relativity,

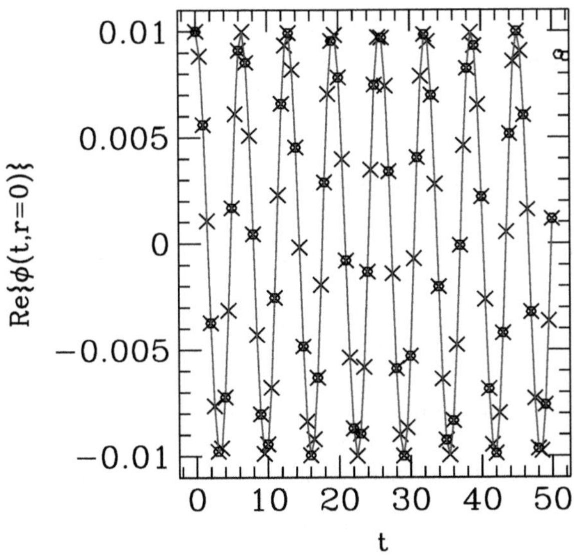

Fig. 7.3 Phase oscillation of the real part of the scalar field at the *center*, up to $t = 50$, for the resolution $\Delta x = 0.25$. The *continuous line* indicates the analytically expected value, the *crosses* show the numerical solution values for the static coordinates, and the *circles* show the same for the harmonic coordinates.

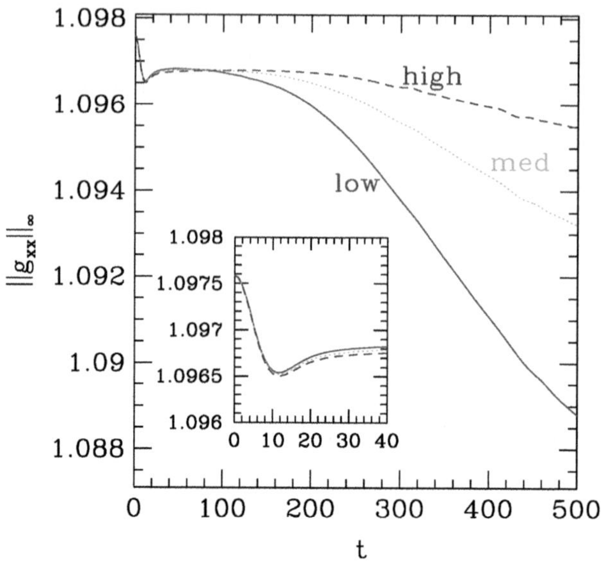

Fig. 7.4 L_∞ norm of g_{xx} versus time for solution obtained employing the harmonic coordinates condition. The figure displays the simulation results for three different base resolutions $\Delta x = \{0.25, 0.375, 0.50\}$. After a short transient behavior, we get convergence toward a constant value.

Maxwell equations must be satisfied in addition to the energy and momentum conservation.

7.2.1 Maxwell equations

Maxwell equations can be written in general covariant form as

$$\nabla_\nu F^{\mu\nu} = 4\pi I^\mu, \tag{7.27}$$

$$\nabla_\nu {}^*F^{\mu\nu} = 0, \tag{7.28}$$

where $F^{\mu\nu}$ is the Maxwell tensor of the electromagnetic field, ${}^*F^{\mu\nu}$ is the Faraday tensor, and I^μ is the electric current four-vector. Since $F^{\mu\nu}$ is antisymmetric, the four-divergence of (7.27) leads to the current conservation

$$\nabla_\mu I^\mu = 0. \tag{7.29}$$

When both the electric and magnetic susceptibilities of the medium vanish, like in vacuum or in a highly ionized plasma, the Faraday tensor is simply the dual of the Maxwell one, that is,

$$
{}^*F^{\mu\nu} = \frac{1}{2}\,\epsilon^{\mu\nu\alpha\beta}\,F_{\alpha\beta}\,, \qquad F^{\mu\nu} = -\frac{1}{2}\,\epsilon^{\mu\nu\alpha\beta}\,{}^*F_{\alpha\beta}, \tag{7.30}
$$

where $\epsilon^{\mu\nu\alpha\beta}$ is the Levi–Civita pseudotensor of the spacetime, which can be written in terms of the four-indices Levi–Civita symbol $\eta^{\mu\nu\alpha\beta}$ as

$$
\epsilon^{\mu\nu\alpha\beta} = \frac{1}{\sqrt{g}}\,\eta^{\mu\nu\alpha\beta}. \tag{7.31}
$$

Allowing for (7.30), the homogeneous equation (7.28) can be written in terms of the Maxwell tensor $F^{\mu\nu}$, namely

$$
\nabla_\mu F_{\nu\alpha} + \nabla_\alpha F_{\mu\nu} + \nabla_\nu F_{\alpha\mu} = 0. \tag{7.32}
$$

In the special relativity case, we can recover in this way the formulae previously discussed in Chap. 2.

There are several ways to represent Maxwell equations. Here we will consider the main ones: the vector potential form and the electromagnetic field system.

7.2.2 Electromagnetic potential

The Maxwell tensor can be written in terms of a four-vector potential A_μ,

$$
F_{\mu\nu} = \nabla_\mu A_\nu - \nabla_\nu A_\mu. \tag{7.33}
$$

The homogeneous Maxwell equation (7.32) holds then identically, allowing for the Bianchi identities of the Riemann tensor. The only non-trivial equation is then the inhomogeneous one,

$$
\nabla_\nu F^{\mu\nu} = \nabla_\nu(\nabla^\mu A^\nu - \nabla^\nu A^\mu) = 4\pi I^\mu\,, \tag{7.34}
$$

which can also be written as

$$
\nabla^\nu\nabla_\nu A_\mu - \nabla_\mu(\nabla_\nu A^\nu) - R^\nu{}_\mu A_\nu = -4\pi I_\mu\,, \tag{7.35}
$$

where we have just changed the ordering of covariant derivatives in (7.34).

Note that, allowing for the antisymmetry of the Maxwell tensor and the conservation equation (7.29), we can write

$$
\nabla_0\,(\nabla_\nu F^{0\nu} - 4\pi\, I^0) + \nabla_k(\nabla_\nu F^{k\nu} - 4\pi\, I^k) = 0 \qquad (k = 1, 2, 3)\,, \tag{7.36}
$$

where latin indices refer to space coordinates. This means that the time component of (7.34), namely

$$\nabla_\nu F^{0\nu} = \nabla_i(\nabla^0 A^i - \nabla^i A^0) = 4\pi I^0, \tag{7.37}$$

is a first integral of the full system (7.34). Equation (7.37) is actually a constraint equation. It follows from (7.36) that, as in the Einstein equations case, this constraint will be preserved by the evolution system provided that it is satisfied by the initial data.

This constraint implies a gauge freedom, meaning that the evolution of electromagnetic potential A_μ is not fully determined. This gauge freedom is evident from the fact that the electromagnetic field expression (7.33) is unchanged by the replacement

$$A_\mu \quad \rightarrow \quad A_\mu + \partial_\mu \Lambda, \tag{7.38}$$

where Λ is an arbitrary function. We can choose this function such that the resulting electromagnetic potential verifies

$$\nabla_\mu A^\mu = 0 \tag{7.39}$$

(Lorentz gauge). Equations (7.35) get then the 'harmonic' form

$$\nabla^\nu \nabla_\nu A_\mu = 8\pi \, [T^\nu{}_\mu - \frac{T}{2}\, \delta^\nu{}_\mu] \, A_\nu - 4\pi I_\mu, \tag{7.40}$$

where we have used Einstein's field equations.

The inhomogeneous (vector) wave equation (7.40) provides a symmetric-hyperbolic evolution system for the electromagnetic potential, for any given spacetime. The Lorentz gauge condition (7.39) is the analogous of the harmonic constraints in the harmonic formulation of Einstein's equations. It will be satisfied at the continuum level, although numerical evolution can produce constraint violations.

We can take these deviations into account by extending the solution space, as we did in Chap. 3 with the gravitational field equations. Let us consider the following extended Maxwell equations:

$$\nabla^\nu (F_{\mu\nu} - g_{\mu\nu}\, \Psi) = 4\pi I_\mu + \kappa\, n_\mu \Psi, \tag{7.41}$$

where we have introduced the 'gauge source' $\Psi \equiv \nabla^\mu A_\mu$ and n_μ is here again the normal to the $t = const$ hypersurfaces. The evolution equation for Ψ is given by the time component of (7.41). The standard Maxwell equations in the Lorentz gauge are recovered when $\Psi = 0$.

Also, by taking the covariant divergence of (7.41) and imposing the conservation of the current, we obtain the following subsidiary evolution equation for the gauge source:

$$\Box\, \Psi = -\kappa\, \nabla^\mu (n_\mu \Psi). \tag{7.42}$$

This equation implies that any gauge-constraint deviation Ψ will propagate with light speed and will be damped by the extra κ terms in (7.41).

7.2.3 The electromagnetic fields

Although the potential vector formulation is very elegant, in most cases it is more intuitive to work with the standard electric and magnetic fields. In this case it is more convenient to decompose the Faraday tensor as

$$F^{\mu\nu} = t^\mu E^\nu - t^\nu E^\mu + \epsilon^{\mu\nu\alpha\beta}\, B_\alpha\, t_\beta \tag{7.43}$$

$$^*F^{\mu\nu} = t^\mu B^\nu - t^\nu B^\mu - \epsilon^{\mu\nu\alpha\beta}\, E_\alpha\, t_\beta \tag{7.44}$$

where $t^\mu = -n^\mu$ is the unit time vector associated to a generic normal observer. The vectors E^μ and B^μ are the electric and magnetic fields measured by this observer. Both fields are purely spatial (i.e., $E^\mu t_\mu = B^\mu t_\mu = 0$).

Note that the electric and magnetic fields depend strongly on the observer. For instance, a moving charged particle produces a measurable electric and magnetic field for most of the observers. On the other hand, an observer comoving with the particle will measure only an electric field, since the particle is at rest with respect to him. There are, however, some scalars constructed from the Maxwell and Faraday tensors, which are then independent of the observer. The simplest ones are just quadratic combinations, namely

$$^*F_{\mu\nu}F^{\mu\nu} = 4E^\mu B_\mu, \quad F_{\mu\nu}F^{\mu\nu} = 2(B^2 - E^2). \tag{7.45}$$

On the other hand, the electric current I^μ can be decomposed into its 3+1 components, namely

$$q \equiv n_\nu I^\nu \ , \qquad J_i = I_i, \tag{7.46}$$

where q and J_i are the charge density and current, respectively, as observed by the normal observer. Current conservation (7.29) can be expressed in the 3+1 form as

$$(\partial_t - \mathcal{L}_\beta)\, q + \nabla_i(\alpha J^i) = \alpha\, q\, trK, \tag{7.47}$$

where the space metric γ_{ij} must be used for covariant derivatives and index raising. We can write it in flux-conservative form, that is,

$$\partial_t\, [\sqrt{\gamma}\, q] + \partial_i\, [\sqrt{\gamma}\, (-\beta^i\, q + \alpha J^i)\,] = 0. \tag{7.48}$$

Note that a prescription for the space components J_i is necessary in order to close the system. This relation, which is commonly known as the Ohm's law, will be discussed in detail later in this chapter.

Let us consider from the very beginning the extended Maxwell equations

$$\nabla_\mu (F^{\mu\nu} + g^{\mu\nu}\Psi) = -4\pi I^\nu - \kappa\, n^\nu \Psi \tag{7.49}$$

$$\nabla_\mu ({}^*F^{\mu\nu} + g^{\mu\nu}\phi) = -\kappa\, n^\nu \phi, \tag{7.50}$$

so that the two extra scalar fields Ψ and ϕ play the same role as the four-vector Z_μ in the Z4 formalism, as discussed in Chap. 3. We will use the techniques described there in order to write down the 3+1 version of (7.49) and (7.50) in terms of the space components of the electromagnetic fields. We can proceed as usual (in normal coordinates first, then extending to the shift case), getting the final 3+1 expressions:

$$(\partial_t - \mathcal{L}_\beta)\, E^i \;-\; \epsilon^{ijk}\nabla_j(\,\alpha B_k\,) + \alpha\,\gamma^{ij}\nabla_j\,\Psi = \alpha\, trK\, E^i - 4\pi\alpha J^{\,i} \tag{7.51}$$

$$(\partial_t - \mathcal{L}_\beta)\, B^i \;+\; \epsilon^{ijk}\nabla_j(\,\alpha E_k\,) + \alpha\,\gamma^{ij}\nabla_j\,\phi = \alpha\, trK\, B^i \tag{7.52}$$

$$(\partial_t - \mathcal{L}_\beta)\,\Psi + \alpha\,\nabla_i E^i = 4\pi\alpha\, q - \alpha\kappa\,\Psi \tag{7.53}$$

$$(\partial_t - \mathcal{L}_\beta)\,\phi + \alpha\,\nabla_i B^i = -\alpha\kappa\,\phi\,. \tag{7.54}$$

Note that the standard Maxwell equations in a curved background are recovered for $\Psi = \phi = 0$. The Ψ and ϕ scalars can then be considered as the normal time integrals of the standard divergence constraints

$$\nabla_i E^i = 4\pi q, \qquad \nabla_i B^i = 0. \tag{7.55}$$

The subsidiary conditions, like (7.42), ensure that the constraints will propagate with light speed and that they will be damped during the evolution.

Another important difference between the standard and the extended Maxwell equations arises when considering the characteristic structure. The spectral decomposition of the system (7.51), (7.52), (7.53), and (7.54), with respect to a direction given by a normalized spatial vector $\mathbf{n}$, belonging to the ordered orthonormal triad $\{\mathbf{l}, \mathbf{m}, \mathbf{n}\}$, provides the following list of eigenfields:

- **Constraint eigenfields**. They involve the extra scalar fields and the longitudinal vector components, namely

$$\Psi \pm E_n, \quad \phi \pm B_n, \tag{7.56}$$

  where the symbol $\{l, m, n\}$ replacing an index means the projection along the corresponding vector.
- **Light eigenfields**. These modes represent the physical electromagnetic waves, which are perpendicular to the direction of propagation $\mathbf{n}$:

$$E_l \pm B_m, \quad E_m \mp B_l. \tag{7.57}$$

The extended evolution system is fully degenerate, so that the local characteristic speed is $-\beta^n \pm \alpha$ in all cases.

7.2.4 The electromagnetic stress–energy tensor

The stress–energy tensor for the electromagnetic fields is given in terms of the Maxwell tensor

$$T_{\mu\nu} = \frac{1}{4\pi}\left[F_{\mu}{}^{\lambda}\, F_{\nu\lambda} - \frac{1}{2}\, g_{\mu\nu}\, F^{\lambda\sigma}F_{\lambda\sigma}\right]. \tag{7.58}$$

After some algebra the 3+1 components of the stress–energy tensor can be obtained in terms of the electric and magnetic fields, that is,

$$4\pi\,\tau = \frac{1}{2}(E^2 + B^2), \quad 4\pi\, S_i = \epsilon_{ijk} E^j B^k, \tag{7.59}$$

$$4\pi\, S_{ij} = -E_i E_j - B_i B_j + \frac{1}{2}\,\gamma_{ij}\,(E^2 + B^2), \tag{7.60}$$

where $E^2 \equiv E^k E_k$ and $B^2 \equiv B^k B_k$. The scalar component τ can be identified with the energy density of the electromagnetic field and the energy flux S_i is the Poynting vector.

The conservation of the stress–energy tensor leads to evolution equations for both the energy density and the Poynting flux. The same equations can be obtained directly from Maxwell equations. Notice, however, that the opposite is not true: Maxwell equations do not follow from energy and momentum conservation. The divergence of the EM stress–energy tensor (7.58), allowing for the Maxwell equations, leads to

$$\nabla_\nu T^{\mu\nu} = -F^{\mu\nu} I_\nu. \tag{7.61}$$

This result is not inconsistent with the conservation of the energy and momentum. When there are only electromagnetic fields without sources (i.e., $I_\nu = 0$) we get strict conservation. The presence of an electric current implies that there is a charged fluid and there can be transfers of energy and momentum between the EM fields and the fluid, so only the total stress–energy tensor (field plus fluid) is conserved. We will see this in detail in the magnetohydrodynamics section.

7.3 Hydrodynamics

Many astrophysical objects in the universe can be described by a fluid approximation, where each fluid element contains a large number of either particles or molecules. In this approach, only averaged thermodynamical quantities are necessary in order to describe these fluid elements, whose dynamics is provided by the conservation of mass, energy, and momentum. In this section we will study the simplest fluid model (i.e. the perfect fluid) which becomes a good approximation for many astrophysical situations.

7.3.1 Perfect fluids

The stress–energy tensor for a perfect fluid (i.e., neglecting non-adiabatic effects like viscosity and heat transfer) is given by

$$T_{\mu\nu} = [\,\rho(1+\epsilon) + p\,]\, u_\mu u_\nu + p\, g_{\mu\nu}. \tag{7.62}$$

Let us explain in detail each quantity in this expression. The rest mass density ρ is the density of the fluid measured by a comoving observer. It contributes to the total energy density of the fluid μ in the local rest frame, which contains also other terms coming from the internal degrees of freedom of the particles,

$$\mu = \rho\,(1+\epsilon), \tag{7.63}$$

where the (specific) internal energy ϵ accounts for the thermal energy, the binding energy, etc. The pressure p is given by an equation of state as a function of the rest mass density and the internal energy. The equation of state characterizes the type of fluid which is being considered. We can also compute the enthalpy h, namely

$$h = \mu + p = \rho\,(1+\epsilon) + p. \tag{7.64}$$

It is important to stress that the set of thermodynamic quantities $\{\rho, \epsilon, p\}$ are all defined in the rest frame of the fluid element, although in general we will use an Eulerian perspective, where the coordinates are not tied to the flow of the fluid. Therefore, we will need the fluid four-velocity u^μ to describe how the fluid moves with respect to the Eulerian observers. The four-velocity follows the usual normalization relation:

$$u^\mu u_\mu = -W^2 + \gamma^{ij}\, u_i u_j = -1, \tag{7.65}$$

where we have introduced here the 3+1 four-velocity components:

$$W \equiv n_\mu u^\mu = \alpha\, u^0, \qquad u_i. \tag{7.66}$$

In general, it is more useful to deal with the three-velocity vector v^i, namely

$$u_i = W v_i, \qquad W = (1 - \gamma^{ij}\, v_i v_j)^{-1/2} , \tag{7.67}$$

so that we can recognize W as the general relativistic Lorentz factor. The set of fluid variables $U = (\rho, \epsilon, p, v_i)$ provides the primitive quantities describing the state of a perfect fluid.

In addition to the conservation of energy and momentum, when there is neither creation nor destruction of particles, the fluid must conserve also the total number of baryons. This law is expressed in terms of the baryon number density $J^\mu = \rho u^\mu$ and is written as

$$\nabla_\mu J^\mu = 0, \tag{7.68}$$

which is just the relativistic generalization of the conservation of mass.

7.3.1.1 Conservation laws

The first law of thermodynamics states that the total energy is preserved. In the case of a simple fluid, we can write it in the form

$$dU = \delta Q - p\, dV, \tag{7.69}$$

where U is the total internal energy and Q is the heat added to the fluid. For reversible processes, we have

$$\delta Q = T dS, \tag{7.70}$$

where S is the entropy of the fluid. All these thermodynamic quantities are defined in the fluid's rest frame. The second law of thermodynamics

$$u^\mu \nabla_\mu S \geq 0 \tag{7.71}$$

states that the entropy can increase, although it is conserved along flow lines when the fluid is in thermal equilibrium. Entropy does change, for instance, when shocks appear due to the genuine non-linearity of the fluid evolution equations. This increase in entropy due to shocks is associated with the transfer of energy of bulk motion into internal energy (i.e., heat).

In our approach, we will consider a fixed reference volume, so that the first law for reversible processes will be written as

$$d\epsilon = T ds + \frac{p}{\rho^2}\, d\rho , \tag{7.72}$$

where ϵ and s are the internal energy and entropy densities, respectively.

In order to capture properly the weak solutions (including shocks) of the non-linear equations, it is important to write them in local conservation law form. We have seen in Chap. 4 how this leads to the required

Rankine–Hugoniot (jump) conditions across discontinuities. Let us start with the simplest equation. Baryon conservation can be written in flux-conservative form by using the standard 3+1 decomposition, namely

$$\partial_t(\sqrt{\gamma}\,D) + \partial_j\,[\,\sqrt{\gamma}\,D\,(\alpha\,v^j - \beta^j)\,] = 0, \tag{7.73}$$

where we have defined the conserved quantity $D \equiv n_\mu J^\mu = \rho W$, which is the baryon mass density measured by the Eulerian observers.

The matter stress–energy tensor conservation (7.62) can be put also in 3+1 form, as we did in Chap. 2. This can actually be expressed as a system of balance laws for the energy and momentum densities, namely

$$\partial_t(\sqrt{\gamma}\,\tau) + \partial_j\,[\,\sqrt{\gamma}\,(\alpha S^j - \beta^j\tau)\,] = \sqrt{\gamma}\,[\,\alpha\,S^{ij}K_{ij} - S^j\partial_j\alpha\,] \tag{7.74}$$

$$\partial_t(\sqrt{\gamma}\,S_i) + \partial_j\,[\,\sqrt{\gamma}\,(\alpha S^j{}_i - \beta^j S_i)\,] = \sqrt{\gamma}\,[\,\alpha\,\Gamma_{jki}S^{jk} + S_j\,\partial_i\beta^j - \tau\,\partial_i\alpha\,], \tag{7.75}$$

where the projections of the stress–energy tensor for the perfect fluid can be written in a compact form by using the fluid velocity and the enthalpy,

$$\tau = hW^2 - p\ , \qquad S_i = hW^2 v_i\ , \qquad S_{ij} = hW^2\,v_i v_j + p\,\gamma_{ij}. \tag{7.76}$$

In summary, the evolution of the flux-conserved quantities $V = (\,D,\,\tau,\,S_i\,)$ comes from the baryon number conservation (7.73) and the energy and momentum conservation (7.74) and (7.75). Note that these quantities are strictly conserved only in the special relativistic case (in cartesian coordinates). The energy–momentum equations contain geometric source terms in curved spacetimes.

An equation of state (EOS) is required in order to recover the physical or primitive variables $U = (\,\rho,\,\epsilon,\,p,\,v_i\,)$ from the conserved quantities. The inversion from conserved to primitive variables can involve transcendental equations, depending on the particular EOS, so this process requires special attention. We will discuss further this point later in this chapter.

7.3.1.2 The characteristic structure

The characteristic structure, as defined by the linearization approach [8], has been studied extensively by Anile [9] in the local rest frame and generalized to any other frame in [10] for the special relativistic case and in [11] for full GR. The long list of eigenvectors, given in these references, does not provide much more insight on the system dynamics, so we will only show here the explicit expression of the eigenvalues. As usual, we shall select a specific space direction and the symbol n replacing an index means the projection along the corresponding unit vector n_i.We get:

- **three material waves**, with propagation speed

$$-\beta^n + \alpha v^n \tag{7.77}$$

- **two acoustic waves**, with propagation speed

$$-\beta^n + \frac{\alpha}{1 - v^2 c_s^2}\,[\,(1 - c_s^2) v^n \pm c_s \sqrt{(1 - v^2)\,[\,(1 - v^2 c_s^2) - (1 - c_s^2) v_n^2\,]}\,]\ , \quad (7.78)$$

where c_s is the sound speed.

The sound speed is the (special relativistic) characteristic speed for a pressure (resp. density) perturbation in the local frame comoving with the fluid. This characteristic speed is obtained by linearizing the evolution equations. It is just a function of the equation of state, namely

$$c_s^2 = \left(\frac{\partial p}{\partial \mu}\right)_s , \quad (7.79)$$

or, in terms of ρ and ϵ derivatives,

$$h\, c_s^2 = \rho \left(\frac{\partial p}{\partial \rho}\right)_\epsilon + \frac{p}{\rho}\left(\frac{\partial p}{\partial \epsilon}\right)_\rho . \quad (7.80)$$

7.3.2 The equation of state

The equation of state (EOS) provides a connection between the microscopic properties of the particles and the thermodynamic quantities of the fluid associated with them. In particular, the equation of state relates the pressure in the fluid with two independent quantities, like the rest mass density and the internal energy density

$$p = p\,(\rho, \epsilon), \quad (7.81)$$

so it provides the matter behavior in a particular thermodynamic state. In general the EOS is calculated from sophisticated nuclear physics models of cold plasma. Although these tabulated EOS are more realistic, they are too complicated for simple calculations, so we will describe some simpler EOS which can be written in closed form.

A very common closed-form EOS is the polytropic one,

$$p = K(s) \rho^{\Gamma}, \quad (7.82)$$

which corresponds to a non-interacting, degenerate matter. The 'polytropic constant' $K(s)$ depends on the entropy and Γ is the adiabatic index (we use capital letters in the GR context to avoid confusion with the space metric determinant). For instance, relativistic ideal fermion gases (such as the degenerate relativistic electron gas in white dwarfs) are reasonably well described by a polytropic EOS with $\Gamma = 4/3$. When the temperature of the star is far below its Fermi temperature, it can be considered to be at $T = 0$ (i.e., a 'cold' star). The degenerate matter dynamics in these configurations can be well modeled as an adiabatic flow with a constant K.

The internal energy ϵ is given by the first law of thermodynamics (7.72) for adiabatic processes ($\delta Q = 0$), which can be integrated to obtain

$$\epsilon = \frac{1}{\Gamma - 1} K \rho^{\Gamma - 1} = \frac{p}{\rho\,(\Gamma - 1)}, \tag{7.83}$$

where we have imposed that the internal energy tends to zero in the limit $\rho \to 0$. The relation (7.83) is the relativistic version of the ideal gas law (Boyle's law). With the adiabatic assumption, (7.82) and (7.83) represent a barotropic fluid where the pressure is just a function of the density. Note, however, that the adiabatic index Γ can depend on the dynamical regime given by ρ and ϵ; stiffer fluids have larger Γ and result in more compact stellar configurations. Comparing the observations of neutron stars with the equilibrium solutions obtained with polytropic EOS, we get a good agreement on masses and radius for $\Gamma = 2$ (although the state-of-art computation of EOS nuclear matter seems to indicate larger values [12]).

One of the most serious drawbacks of the polytropic EOS is that, although (7.82) is a good approximation for a 'cold' star, there are processes, like the merger of stars or accretion from a disk, which can increase enormously the temperature and a simple polytrope will not provide a physical description. For those cases, it is usual to discard (7.82) and consider just (7.83) as the ideal gas EOS in the form

$$p = (\Gamma - 1)\,\rho\,\epsilon\,, \tag{7.84}$$

which allows fluid heating due to shocks.

A more realistic EOS in closed form can be obtained by a combination of the polytropic EOS to describe the cold part and an ideal EOS for the thermal one, allowing for heating due to the shocks. The hybrid EOS is given by

$$p = K\rho^{\Gamma} + (\Gamma_{\rm th} - 1)\,\rho\,\epsilon_{\rm th}, \tag{7.85}$$

with an adiabatic thermal index $\Gamma_{\rm th}$ that can be different from the adiabatic cold index Γ. The internal energy can be split into a thermal and a cold part,

$$\epsilon = \epsilon_{\rm th} + \epsilon_{\rm cold}. \tag{7.86}$$

The total internal energy ϵ can be obtained from the evolution of the conserved quantities, while the cold part is described by (7.83), leading to the explicit expression,

$$p = K \frac{\Gamma - \Gamma_{\rm th}}{\Gamma - 1} \rho^{\Gamma} + (\Gamma_{\rm th} - 1)\,\rho\,\epsilon\,. \tag{7.87}$$

Notice that this approach can be generalized in order to use a collection of continuous piecewise polytropes in the hybrid EOS, allowing an accurate matching with any realistic tabulated EOS.

7.3.2.1 The transformation from conserved to primitive quantities

As was commented previously, the conversion from conserved quantities to primitive ones is a complicated process which depends on the particular form of the EOS. In simple cases it involves solving fourth-order algebraic equations, but it becomes a trascendental equation in the generic case. For efficiency and generality, it is usually solved numerically, by using for instance a Newton–Raphson method [5]. Here we detail the inversion for the hybrid EOS (7.85), since it reduces to either the polytropic or the ideal gas EOS for suitable choices of $(K,\, \Gamma_{\rm th})$. Let us remind here the definition of the conserved quantities:

$$D = W\rho \quad , \quad S_i = hW^2 v_i \quad , \quad \tau = hW^2 - p. \tag{7.88}$$

We will solve for the combination $x \equiv hW^2$. We can express the speed in terms of the momentum S_i, getting the following relationships:

$$v_i = S_i/x \;, \qquad W^2 = \frac{x^2}{x^2 - S^i S_i}\;. \tag{7.89}$$

We can start by using the hybrid EOS (7.85) for expressing the enthalpy as

$$h = \rho + \frac{\Gamma}{\Gamma - 1} K\rho^{\Gamma} + \Gamma_{\rm th}\,\rho\,\epsilon_{\rm th} \tag{7.90}$$

and repeat the same process for the energy density in (7.88), namely

$$\tau = \rho W^2 + [\frac{W^2 \Gamma}{\Gamma - 1} - 1]\, K\rho^{\Gamma} + [1 + \Gamma_{\rm th}(W^2 - 1)]\,\rho\,\epsilon_{\rm th}\;. \tag{7.91}$$

Now, the non-polytropic internal energy term $\rho\,\epsilon_{\rm th}$ can be eliminated, getting the final equation:

$$\tau = [1 - \frac{(\Gamma_{\rm th} - 1)}{\Gamma_{\rm th} W^2}]\, x + \frac{\Gamma_{\rm th} - 1}{\Gamma_{\rm th} W} D + \frac{\Gamma_{\rm th} - \Gamma}{\Gamma_{\rm th}(\Gamma - 1)} K \left(\frac{D}{W}\right)^{\Gamma}\;. \tag{7.92}$$

The physical solution is obtained for a value $x = x_{\rm sol}$ such that (7.92) holds true. One can start with an initial guess for x, usually the value obtained in the previous timestep and proceed by iteration, until the required condition (7.92) is satisfied.

7.3.3 Neutron stars

Neutron stars can be modeled by a perfect fluid with a stiff equation of state. As in the case of the boson stars, we consider spherically symmetric static solutions of the Einstein equations coupled to the hydrodynamic ones. Let us consider the line element in Schwarzschild coordinates

$$ds^2 = -\alpha^2(r)\,dt^2 + a^2(r)\,dr^2 + r^2\,d\Omega^2. \tag{7.93}$$

The mass function (7.19) in this case takes the simple expression

$$m(r) = \frac{r}{2}(1 - 1/a^2), \tag{7.94}$$

which accounts for the mass at a given radius.

Einstein's field equations can be reduced in this case (where both the fluid and the geometry are spherically symmetric and static) to the Tolman–Oppenheimer–Volkoff (TOV) equilibrium equations,

$$m' = 4\pi r^2 \mu \tag{7.95}$$

$$p' = -\frac{(\mu + p)(m + 4\pi r^3 p)}{r(r - 2m)} \tag{7.96}$$

$$(ln\alpha)' = \frac{(m + 4\pi r^3 p)}{r(r - 2m)}, \tag{7.97}$$

which require the prescription of some EOS in order to obtain the matter density ρ. As discussed before, the polytropic EOS

$$p = K\rho^{\Gamma} \tag{7.98}$$

provides a good description for a cold star with $\Gamma = 2$. In geometrized units ($G = c = 1$), the constant K sets the length scale of the system, so we can set $K = 1$ to get adimensional units. With these choices we can integrate the TOV equations from the origin up to the surface of the star, which is defined as the radius where the pressure vanishes. The metric functions are continued past the star's radius by matching to the Schwarzschild solution with a mass given by

$$m(r) = M_{\text{star}} \qquad (r \geq R_{\text{star}})\ . \tag{7.99}$$

The solutions are very similar to those obtained for boson stars. For a given value of the central density $\rho_c = \rho_o(0)$ these equations can be solved

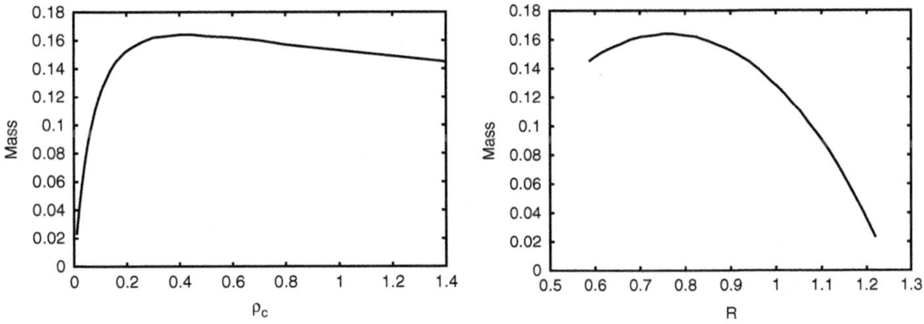

Fig. 7.5 Total mass of the neutron star as a function of either the central density value (*left panel*) or the radius (*right panel*). The maximum stable mass, for $K = 1$ and $\Gamma = 2$, is at $M = 0.164$. The stable branch is at the *left (resp. right) side* of the maximum in the *left (resp. right) panel.*

by integrating from $r = 0$ outward using a standard ODE integrator. The regularity conditions at the origin require taking

$$m(0) = 0 \ , \qquad (ln\alpha)' = 0 \ \Leftrightarrow \ p' = 0 \ . \tag{7.100}$$

The lapse needs here again to be rescaled after the integration in order to match the Schwarzschild solution at the surface of the star. Once the solution is computed in this coordinate system, a change of coordinates is performed to isotropic ones:

$$ds^2 = \alpha^2 \left(\tilde{r}\right) dt^2 + \psi^4 \left(\tilde{r}\right) \left(d\tilde{r}^2 + \tilde{r}^2 d\Omega^2\right) \tag{7.101}$$

(we use here $\tilde{r}$ instead of ρ to avoid confusion with the matter density).

The diagrams of the neutron stars are very similar qualitatively to the boson star ones, as we can see in Fig. 7.5. The matter density takes here the role of the scalar field in Fig. 7.3. On the left panel, we show the mass as a function of the central density ρ_c. The stable branch is located here on the left of the allowed maximum mass, satisfying

$$\frac{\partial M}{\partial \rho_c} \geq 0 \ . \tag{7.102}$$

The neutron star configurations on this branch are stable against small perturbations, while the ones on the opposite branch are unstable. The final state can be either a neutron star in the stable branch or a black hole.

The profile of the neutron star used for evolution in the next section is shown here as an example. The density and the conformal factor are shown in Fig. 7.6.

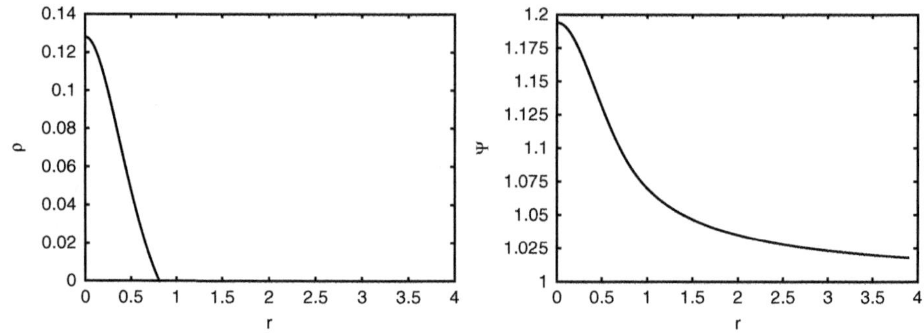

Fig. 7.6 Profiles of the density and the conformal factor, in isotropic coordinates, for a neutron star in the stable branch.

7.3.3.1 Evolution of single neutron star

A standard test involving the TOV solution is to accurately reproduce the known radial oscillation modes of the star. While the TOV solution is spherically symmetric and static, discretization effects act as small perturbations that excite the normal modes of the star.

The initial data for this test consist of a $\Gamma = 2$ polytrope, which can be obtained by solving the system introduced in detail in the previous section. The star, in the geometrized units with $K = 1$, has a mass of $M = 0.14$, an area radius $R = 0.958$, and central rest mass density $\rho_c = 0.128$. We evolve the data in a dynamic spacetime for different resolutions. Figure 7.7 shows ρ_c plotted as a function of time for three resolutions: 32, 64, and 128 points across the star. As expected, both the amplitude oscillations and the overall drift in ρ_c actually converge with resolution. This is important both as a code test and as an indication of the resolution necessary to capture some dynamics of stellar interiors. The data in Fig. 7.7 were generated using a third-order accurate method. We found that first- and second-order methods were more diffusive, resulting in larger drifts in ρ_c. Consequently, it was more difficult to reproduce the radial pulsation modes of the star using these lower order methods. Owing to the computational costs of these simulations, the higher resolution runs were not evolved up to the same end time.

To confirm that the code reproduces the expected physical behavior, we examine the radial pulsations of the star. The modes are calculated from the oscillations in ρ_c, and the extracted frequencies (obtained by performing the Fourier transform) are shown in Table 7.1. These oscillation modes can be compared to the known radial perturbation modes [13], and the frequencies are in excellent agreement. These validations are a stringent test of our computational methods and give us considerable confidence that our code accurately reproduces the physics of these systems.

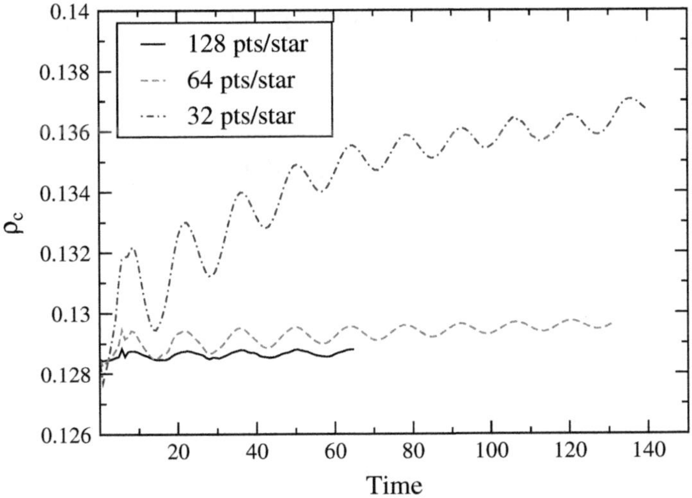

Fig. 7.7 This figure shows oscillations in the central rest mass density ρ_c for the evolution of a single TOV star at three different resolutions: 32, 64, and 128 points across the star. The initial data are for a star of mass $M = 0.14$, area radius 0.958, central rest mass density $\rho_c = 0.128$, $\Gamma = 2$, and $K = 1$. While ρ_c increases noticeably for the coarsest resolution run, it eventually stabilizes at a higher value, giving a stable configuration.

Table 7.1 Comparison of small radial pulsation frequencies for a neutron star, evolved using the 3D GRHD code, with the linear perturbation modes [13]. The polytrope is constructed for $\Gamma = 2$ and $K = 1$.

Mode	3D GRHD code (kHz)	Perturbation results (kHz)	Relative difference (%)
F	14.01	14.42	2.88
H1	39.59	39.55	0.1
H2	59.89	59.16	1.2
H3	76.94	77.76	1.1

7.4 Magnetohydrodynamics

An effective description of a fluid in the presence of electromagnetic fields can be made by considering three different sets of equations governing, respectively, the electromagnetic fields, the fluid variables, and the coupling between them. In particular, the electromagnetic part can be described via the Maxwell equations, while the conservation of mass, energy and momentum can be used to express the evolution of the fluid variables. Finally, Ohm's law, whose exact form depends on the microscopic properties of the fluid, expresses the coupling between the electromagnetic fields and the fluid variables. In what follows we consider these three sets of equations and discuss how they lead to the resistive MHD description, which can be reduced either to the ideal MHD or to the force-free limit.

7.4.1 The MHD evolution equations

Let us assume that the perfect fluid and the electromagnetic fields are minimally coupled, so there are no mixed terms and the total stress–energy tensor can be obtained just by addition

$$T^{\mu\nu} = T^{\mu\nu}_{(\text{fluid})} + T^{\mu\nu}_{(\text{em})}, \tag{7.103}$$

which can be written explicitly as

$$T_{\mu\nu} = [\,\rho(1+\epsilon) + p\,]\, u_\mu u_\nu + p\, g_{\mu\nu} + \frac{1}{4\pi}\, [F_\mu{}^\lambda\, F_{\nu\lambda} - \frac{1}{2}\, g_{\mu\nu}\; F^{\lambda\alpha} F_{\lambda\alpha}]. \tag{7.104}$$

The physical system is now described by the hydrodynamic and the electromagnetic physical fields $U = (\,\rho,\, p,\, \epsilon,\, v_i,\, E_i,\, B_i,\, q\,)$. The evolution equations for the electromagnetic fields are given by the Maxwell equations (7.27) and (7.32) and the conservation of charge (7.29), while the fluid fields are still governed by the conservation of the total energy–momentum (7.1) and baryonic number (7.68). Allowing for the result (7.61) for the electromagnetic part, we get

$$\nabla_\nu T^{\mu\nu} = 0 \quad \rightarrow \quad \nabla_\nu T^{\mu\nu}_{(\text{fluid})} = -\nabla_\nu T^{\mu\nu}_{(\text{em})} = F^{\mu\nu} I_\nu. \tag{7.105}$$

In this form it is evident that, as far as the current I_ν does not contain derivatives of the physical fields, the electromagnetic fields and the fluid evolutions are decoupled in the principal part: they interact only through source terms. This has some implications:

- The fluid contains the charges which produce the current I_ν, so the conducting fluid is the source of the Maxwell equations. An effective relation between the charged fluid and the current (the Ohm's law) will be discussed below.
- The EM field produces a (4D) Lorentz force $F^{\mu\nu} I_\nu$ on the charges of the fluid. The time component gives an extra source (the Joule heating) to the energy equation, while the space components (the Lorentz force) contribute to the fluid motion through the momentum equation.

In addition to the stress–energy conservation, it is also convenient to write the Maxwell equations in balance law form. The full system of equations describing a conducting fluid in the presence of electromagnetic fields is then

$$
\begin{aligned}
\partial_t(\sqrt{\gamma}\,B^i) + \partial_k[\,-\beta^k\sqrt{\gamma}\,B^i + \alpha\epsilon^{ikj}\sqrt{\gamma}\,E_j\,] = \\
-\sqrt{\gamma}\,B^k(\partial_k\beta^i) - \alpha\sqrt{\gamma}\,\gamma^{ij}\partial_j\phi \qquad (7.106)\\
\partial_t(\sqrt{\gamma}\,E^i) + \partial_k[\,-\beta^k\sqrt{\gamma}\,E^i - \alpha\epsilon^{ikj}\sqrt{\gamma}\,B_j\,] = \\
-\sqrt{\gamma}\,E^k(\partial_k\beta^i) - \alpha\sqrt{\gamma}\,\gamma^{ij}\partial_j\Psi - 4\pi\alpha\sqrt{\gamma}\,J^i \qquad (7.107)\\
\partial_t\phi + \partial_k[\,-\beta^k\phi + \alpha B^k\,] = \\
-\phi\,(\partial_k\beta^k) + B^k(\partial_k\alpha) - \alpha\Gamma^i_{ki}B^k - \alpha\kappa\phi \qquad (7.108)\\
\partial_t\Psi + \partial_k[\,-\beta^k\Psi + \alpha E^k\,] = \\
-\Psi\,(\partial_k\beta^k) + E^k(\partial_k\alpha) - \alpha\Gamma^i_{ki}E^k + 4\pi\alpha q - \alpha\kappa\Psi \qquad (7.109)\\
\partial_t(\sqrt{\gamma}\,q) + \partial_k[\,-\beta^k\sqrt{\gamma}\,q + \alpha\sqrt{\gamma}\,J^k\,] = 0 \qquad (7.110)\\
\partial_t(\sqrt{\gamma}\,D) + \partial_k[\,\sqrt{\gamma}\,D\,(\alpha v^k - \beta^k)\,] = 0 \qquad (7.111)\\
\partial_t(\sqrt{\gamma}\,\tau) + \partial_k[\,\sqrt{\gamma}\,(\alpha S^k - \beta^k\tau)\,] = \sqrt{\gamma}\,[\,\alpha S^{ij}K_{ij} - S^j\partial_j\alpha\,] \qquad (7.112)\\
\partial_t(\sqrt{\gamma}\,S_i) + \partial_k[\,\sqrt{\gamma}\,(\alpha S^k{}_i - \beta^k S_i)\,] = \sqrt{\gamma}\,[\,\alpha\Gamma^j_{ik}S^k{}_j + S_j\,\partial_i\beta^j - \tau\partial_i\alpha\,], \qquad (7.113)
\end{aligned}
$$

where the components of the total stress–energy tensor are computed as the linear superposition of the fluid and Maxwell ones, namely

$$
\begin{aligned}
\tau &= hW^2 - p + \frac{1}{8\pi}\,(E^2 + B^2), \qquad (7.114)\\
S_i &= hW^2 v_i + \frac{1}{4\pi}\,\epsilon_{ijk}\,E^j B^k, \qquad (7.115)\\
S_{ij} &= p\,\gamma_{ij} + hW^2\,v_i v_j + \frac{1}{4\pi}\,[-E_iE_j - B_iB_j + \frac{1}{2}\,\gamma_{ij}\,(E^2 + B^2)]. \qquad (7.116)
\end{aligned}
$$

It is interesting to stress that the conversion procedure from conserved to primitive variables described in Sect. 7.3.2.1 can still be used by subtracting the electromagnetic part, i.e.,

$$
\begin{aligned}
\tau &\to \tau - \frac{1}{8\pi}\,(E^2 + B^2), \qquad (7.117)\\
S_i &\to S_i - \frac{1}{4\pi}\,\epsilon_{ijk}\,E^j B^k, \qquad (7.118)
\end{aligned}
$$

which is already known from the Maxwell equations.

7.4.2 Generalized Ohm's law

As mentioned above, Maxwell equations are coupled to the fluid ones by means of the electromagnetic current four-vector I^μ, its explicit form depending in general on the electromagnetic fields and on the local fluid properties, measured in the local comoving frame. For this reason, it is convenient to introduce the electric and magnetic fields measured by an observer moving with the fluid speed u^μ, namely

$$e^\mu \equiv F^{\mu\nu} u_\nu \quad , \quad b^\mu \equiv {}^*F^{\mu\nu} u_\nu \tag{7.119}$$

(here again each one has only three independent components since $e^\mu u_\mu = b^\mu u_\mu = 0$).

Allowing for (7.43) and (7.44), these comoving fields can be easily related with the ones measured by a standard Eulerian observer, namely

$$e^\mu = W E^\mu - n^\mu (E^\nu u_\nu) - \epsilon^{\mu\nu\alpha\beta} u_\nu B_\alpha n_\beta \tag{7.120}$$
$$b^\mu = W B^\mu - n^\mu (B^\nu u_\nu) + \epsilon^{\mu\nu\alpha\beta} u_\nu E_\alpha n_\beta \ . \tag{7.121}$$

Their 3+1 components can be obtained in a straightforward way (the normal coordinates result is also valid for the shift case):

$$n_\mu e^\mu = W v^k E_k \ , \qquad e_i = W\,(E_i + \epsilon_{ijk} v^j B^k), \tag{7.122}$$
$$n_\mu b^\mu = W v^k B_k \ , \qquad b_i = W\,(B_i - \epsilon_{ijk} v^j E^k) \ . \tag{7.123}$$

In the same spirit, we can also decompose the electromagnetic four-current as

$$I^\mu = u^\mu \tilde{q} + j^{\,\mu}, \tag{7.124}$$

where $\tilde{q}$ and $j^{\,\mu}$ are the charge density and the electric current measured by the comoving observer, respectively, and $j^{\,\mu} u_\mu = 0$. Their 3+1 components can also be obtained in the same way:

$$q = W\tilde{q} + j^{\,\mu} n_\mu, \qquad J_i = j_{\,i} + \tilde{q}\, W v_i = q\, v_i + j_{\,i} - (j^{\,\mu} n_\mu)\, v_i, \tag{7.125}$$

so that everything is fully determined by the prescription of the current j in the local comoving frame.

A standard prescription, known as Ohm's law, is to consider the current to be proportional to the Lorentz force acting on charged particles, that is, a linear relation between $j^{\,\mu}$ and e^μ,

$$j^{\,\mu} = \sigma^{\mu\nu} e_\nu, \tag{7.126}$$

where $\sigma^{\mu\nu}$ is the electrical conductivity of the medium. This conductivity can be calculated in the collision-time approximation [14], and the result may be written in covariant four-tensor form [15]:

$$\sigma^{\mu\nu} = \sigma\,(g^{\mu\nu} + \xi^2 b^\mu b^\nu + \xi\, \epsilon^{\mu\nu\alpha\beta} u_\alpha b_\beta). \tag{7.127}$$

The first term leads to the well-known isotropic (scalar) Ohm's law, while the other two represents the anisotropies due to the presence of magnetic fields. The coefficients are given by

$$\xi = e\tau/m \ , \qquad \sigma = \frac{n_e e\, \xi}{1 + \xi^2 b^2}, \tag{7.128}$$

where τ is the collision time, n_e is the electron density, e and m are the electron's charge and mass, and $b^2 = b^\mu b_\mu$.

It is important to remember that in deriving expression (7.126) for Ohm's law we are implicitly assuming that the collision frequency of the constituent particles of our fluid is much larger that the typical oscillation frequency of the plasma. Stated differently, the timescale for the electrons and ions to come into equilibrium is much shorter than any other timescale in the problem, so that no charge separation is possible and the fluid is globally neutral. This assumption is a key aspect of the MHD approximation. Moreover, since the ratio $e/m >> 1$ (we are using geometrized units), we can assume that in general the second term in (7.127) is much larger than the third one, which will be neglected in what follows.

Now we are in a position of getting an explicit expression for the current J_i in terms of fields measured by an Eulerian observer. Let us start from the corresponding expression in the local comoving frame, namely

$$j_i = \sigma \left[e_i + \xi^2 \left(E^k B_k \right) b_i \right], \tag{7.129}$$

where we have used

$$e^\mu b_\mu = E^\mu B_\mu = E^k B_k. \tag{7.130}$$

Allowing for (7.122), (7.123), and (7.125), we get the final expression

$$\begin{aligned} J_i = q\, v_i &+ W\sigma \left[E_i + \epsilon_{ijk} v^j B^k - (v_k E^k) v_i \right] \\ &+ W\sigma\, \xi^2 (E^k B_k) \left[B_i - \epsilon_{ijk} v^j E^k - (v_k B^k)\, v_i \right]. \end{aligned} \tag{7.131}$$

Note that the conservation of the electric charge (7.110) provides the evolution equation for the charge density q. Ohm's law (7.131) provides the missing prescription for the (spatial) conduction current J.

7.4.2.1 The high-conductivity limit

The general system of (resistive) MHD equations raises a delicate issue when the conductivity in the plasma undergoes very large spatial variations. In the regions with high conductivity, in fact, the system will evolve on timescales which are very different from those in the low-conductivity region. Mathematically, therefore, the problem can be regarded as a hyperbolic system with stiff relaxation terms which requires special care to capture the dynamics in a stable and accurate manner.

A prototypical hyperbolic equation with relaxation is given by

$$\partial_t \mathbf{u} = H(\mathbf{u}, \partial \mathbf{u}) + \frac{1}{\epsilon}\, R(\mathbf{u}), \tag{7.132}$$

where $\epsilon > 0$ is the *relaxation time* (not necessarily constant either in space or in time), H gives rise to a quasilinear system of equations (i.e., H depends linearly on first derivatives of $\mathbf{u}$), and R does not contain derivatives of $\mathbf{u}$.

In the limit $\epsilon \to \infty$ (corresponding to the resistive MHD equations for the case of vanishing conductivity) the system is hyperbolic with propagation speeds bounded by the hyperbolic speed c_h. This maximum bound, together with the length scale L of the system, defines a characteristic timescale $\tau_h \equiv L/c_h$ for the hyperbolic part.

In the opposite limit $\epsilon \to 0$ (corresponding to the case of infinite conductivity), the system is instead said to be *stiff*, since the timescale ϵ of the relaxation (or stiff) term $R(\mathbf{u})$ is in general much shorter than the timescale τ_h of the hyperbolic part. In such a limit, the stability of an explicit scheme is only achieved with a timestep size $\Delta t \leq \epsilon$. This requirement is certainly more restrictive than the Courant–Friedrichs–Lewy (CFL) stability condition $\Delta t \leq \Delta x/c_h$ for the hyperbolic part and makes an explicit integration impractical. The development of efficient numerical schemes for such systems is challenging, since in many applications the relaxation time can vary by several orders of magnitude across the computational domain and, more importantly, becoming much shorter than the timescale determined by the hyperbolic speed c_h.

When faced with this issue several strategies can be adopted. The most straightforward one is to consider only the stiff limit $\epsilon \to 0$, where the system is well approximated by a suitable reduced set of conservation laws called 'equilibrium system' [16] such that

$$R(\bar{\mathbf{u}}) = \mathbf{0}\,, \tag{7.133}$$

$$\partial_t \bar{\mathbf{u}} = \mathbf{G}(\bar{\mathbf{u}},\, \partial \bar{\mathbf{u}}), \tag{7.134}$$

where $\bar{\mathbf{u}}$ is a reduced set of variables. This approach can be followed if the resulting system is also hyperbolic. This is precisely the case in the resistive MHD equations for infinite conductivity. In this case, depending on which component of the conductivity is infinite, the equations reduce to those of either the ideal MHD or the force-free case, which are both described indeed by an 'equilibrium system' in which the conductivity does not appear in the equations.

7.4.3 Ideal MHD

Let us assume that there is a large amount of free electrons, so $(n_e\, e\, \xi)$ is large, while at the same time $\xi\, |b| << 1$ (i.e., the collision time τ is small compared with the electron Larmor period). In this limit the conductivity (7.127) reduces to

$$\sigma^{\mu\nu} \approx \sigma_I g^{\mu\nu}, \qquad \sigma_I = n_e\, e^2 \tau / m \ , \tag{7.135}$$

which describes accurately an isotropic highly conducting hot fluid when the magnetic field is not too large, since the electron density is high and, although the collision time is short, the product $(n_e\,\tau)$ is not small.

The well-known ideal MHD limit of Ohm's law can be obtained by requiring the current to be finite even in the limit of infinite conductivity $(\sigma_I \to \infty)$. This implies that the electric field e^μ measured by the comoving observer must vanish. Allowing for (7.122), we get the well-known ideal MHD condition

$$E^i = -\epsilon^{ijk} v_j B_k \ , \tag{7.136}$$

stating that in this limit the electric field is orthogonal to both **B** and **v**. Such a condition also expresses the fact that in ideal MHD the electric field is not an independent variable, since it can be computed via a simple algebraic relation from the fluid velocity and the magnetic vector field.

Note that the current j becomes undetermined in the relation (7.126), but the redundant Maxwell equations (7.51) and (7.53) may be used to compute the four-current I_μ directly (assuming the vanishing of Ψ). The evolution equations for the magnetic field can be obtained by substituting (7.136) in (7.52) and (7.54), namely

$$\begin{aligned}
\partial_t(\sqrt{\gamma}\, B^i) + \partial_k[\sqrt{\gamma}\, \{(\alpha v^k - \beta^k) B^i - \alpha v^i B^k + \alpha \gamma^{ki} \phi\}] \\
= \sqrt{\gamma}\, [-B^k \partial_k \beta^i + \phi\, \gamma^{ik} (\partial_k \alpha + \Gamma^j_{jk})] \qquad (7.137)\\
\partial_t \phi + \partial_i(-\beta^i \phi + \alpha B^i) = B^i \partial_i \alpha - \alpha B^i \Gamma^k_{ki} - \phi\, \partial_i \beta^i - \kappa \alpha \phi. \qquad (7.138)
\end{aligned}$$

The electric field can be removed also from the set of conserved quantities, in order to convert into primitive quantities in a more robust way, that is,

$$\begin{aligned}
\tau &= h W^2 + B^2 - p - \frac{1}{2}\, [\, (B^k v_k)^2 + \frac{B^2}{W^2}\,] \ , && (7.139)\\
S_i &= [\, h W^2 + B^2\,]\, v_i - (B^k v_k)\, B_i \ . && (7.140)
\end{aligned}$$

The transformation from conserved to primitive is similar to that for an unmagnetized fluid described before in this chapter. The first step is to note that the scalar product $\mathbf{B} \cdot \mathbf{v}$ can be obtained by contracting (7.140) with **B**, namely

$$v_i B^i = \frac{S_i B^i}{h W^2}. \tag{7.141}$$

Using this relation, the scalar product $S^i S_i$ can be solved for the Lorentz factor again, obtaining

$$W^2 = [\,1 - \frac{x^2 S^2 + (2x + B^2)(S_i B^i)^2}{x^2 (x + B^2)^2}\,]^{-1}. \tag{7.142}$$

Now we can still use the final expression (7.92), with the substitution

$$\tau \to \tau - (1 - \frac{1}{2W^2})\, B^2 + \frac{(B^i S_i)^2}{2x^2} \; . \tag{7.143}$$

7.4.3.1 Characteristic structure of the ideal MHD

The characteristic structure of the relativistic ideal MHD equations is much more involved than that of relativistic hydrodynamics. As in classical MHD, there are seven physical waves: two Alfven waves, two fast and two slow magnetosonic waves, and one entropy wave. The characteristic structure of these equations in the fully relativistic case was studied by Anile [9]. It was found that only the entropic waves and the Alfven waves can be explicitly written in closed form, while the other four velocities are found by solving a quartic polynomial:

- **One entropic wave**, with speed

$$\lambda_{\text{entropic}} = -\beta^n + \alpha v^n \; , \tag{7.144}$$

which reduces to a material wave for vanishing magnetic field.
- **Two Alfven waves**, with speeds

$$\lambda^{\pm}_{\text{Alfven}} = -\beta^n + \alpha v^n - \frac{B^n}{hW^2 + B^2}\left[B^k v_k \pm \sqrt{h(B^k v_k)^2 + \frac{B^2}{W^2}} \right], \tag{7.145}$$

which disappear for vanishing magnetic field.
- **Four magnetosonic waves**: the two solutions with maximum and minimum speeds are called fast magnetosonic waves, and slow magnetosonic waves the two solutions in between. Their speeds are given by the solution of the following quartic characteristic equation (which can be obtained numerically):

$$\begin{aligned} 0 &= hW^4(1 - c_s^2)(\alpha v^n - \beta^n - \lambda)^4 \\ &+ [(\beta^n + \lambda)^2 - \alpha^2]\,[\,(\alpha v^n - \beta^n - \lambda)^2 (hW^2 c_s^2 + B^2 + (B^k v_k)^2 W^2\,) \\ &\qquad - c_s^2\,(\,W(B^k v_k)(\alpha v^n - \beta^n - \lambda) + \alpha \frac{B^n}{W}\,)^2\,]. \end{aligned} \tag{7.146}$$

When $B = 0$ the fast magnetosonic modes reduce to the acoustic waves, while the slow ones reduce to material waves.

The propagation speeds of these seven waves, which are displayed in Fig. 7.8, can be ordered as follows:

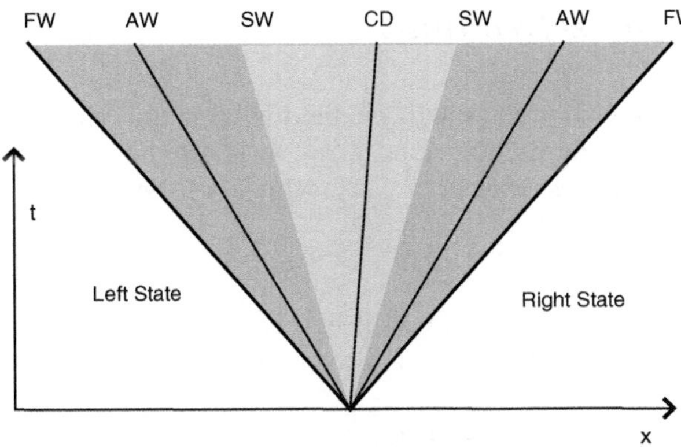

Fig. 7.8 Characteristic structure of the ideal MHD system. It is composed of the four magnetosonic waves – two fast waves (FW) and two slow waves (SW) – two Alfven waves (AW), and a contact discontinuity (CD) given by an entropic wave.

$$\lambda^{+}_{\text{fast}} \geq \lambda^{+}_{\text{Alfven}} \geq \lambda^{+}_{\text{slow}} \geq \lambda_{\text{entropic}} \geq \lambda^{-}_{\text{slow}} \geq \lambda^{-}_{\text{Alfven}} \geq \lambda^{-}_{\text{fast}} \,,$$

defining in this way the domain of dependence of the GRMHD system.

A very useful upper bound for fast waves (which have the maximum speed) can be found by considering the degenerate case of normal propagation [17]. In that case there is an analytical expression for the two fast magnetosonic waves, namely

$$\lambda^{\pm}_{\text{fast}} = -\beta^n + \frac{\alpha}{1 - v^2 a^2} \times \left[(1 - a^2) v^n \pm \sqrt{a^2 (1 - v^2) [(1 - v^2 a^2) - (1 - a^2) v_n^2]} \right] , \quad (7.147)$$

where

$$a^2 = c_s^2 + c_a^2 - c_s^2 c_a^2. \quad (7.148)$$

The sound speed c_s was already defined in (7.80), while the Alfven speed c_a can be written in terms of the comoving magnetic four-vector b^μ:

$$c_a^2 = \frac{b^2}{h + b^2} \,, \qquad b^2 = B^2 - E^2 = B^2/W^2 + (v_k B^k)^2, \quad (7.149)$$

where we have used that, as far as $e^\mu = 0$ in the ideal MHD, the b^2 value coincides with the second invariant in (7.45), which can be computed as well in the frame of the Eulerian observers.

7.4.4 The force-free limit

Let us assume now that the magnetic field is large and the fluid is not so hot (i.e., the collision time is large), so $\xi^2 b^2 \gg 1$. In this case the second term dominates over the first one in (7.127), which can be reduced to

$$\sigma^{\mu\nu} \approx \sigma_B \, b^\mu b^\nu \qquad \sigma_B = n_e e \, \xi / b^2 \,, \tag{7.150}$$

which approximates reasonably well highly conducting cold fluids with low densities. In this case, the Lorentz force computed in terms of the comoving fields is just

$$F^{\mu\nu} I_\nu = \tilde{q} e^\mu + \sigma_B u^\mu (b_\nu e^\nu)^2 \,. \tag{7.151}$$

The analogous of the ideal MHD limit can be obtained by requiring the current to be finite even in the limit of infinite conductivity ($\sigma_B \rightarrow \infty$). Allowing for (7.150), this implies that the electric and magnetic fields must be perpendicular, that is,

$$b^\mu e_\mu = B^\mu E_\mu = \gamma^{ij} \, B_i E_j = 0 \,. \tag{7.152}$$

This condition amounts to the vanishing of the second term in the Lorentz force (7.151), namely

$$F^{\mu\nu} I_\nu = \tilde{q} e^\mu . \tag{7.153}$$

Since the charge density and the electric field in the comoving frame are usually small, we can assume that in general the Lorentz force will be neglectible in this limit. This is the reason why it is called force-free limit, since the Lorentz force is very small.

The force-free limit is usually introduced in a different way. In the magnetospheres of either neutron stars or black holes the density is so low that even moderate magnetic field stresses will dominate over the pressure gradients. This can be modeled by assuming that the stress–energy tensor is dominated by the electromagnetic part:

$$T^{\mu\nu} \approx T^{\mu\nu}_{(\mathrm{em})} \,, \tag{7.154}$$

so that its conservation law implies that the Lorentz force must be negligible:

$$\nabla_\nu T^{\mu\nu}_{(\mathrm{em})} = -F^{\mu\nu} I_\nu \approx 0. \tag{7.155}$$

Let us study in more detail this last relation. Allowing for the electromagnetic tensor decomposition (7.43), the spatial projection of the Lorentz force, relative to the Eulerian observers, is just

$$q E_i + \epsilon_{ijk} J^j B^k = 0 \,, \tag{7.156}$$

which leads consistently to the force-free relation (7.152). This is an indication of how the force-free equations are recovered from an infinite conductivity limit, in a similar way to the ideal MHD.

In this case, the magnetic field evolution is given by the standard Maxwell equations. On the other hand, the electric field cannot be computed properly because of the indetermined current. Instead, we can use the conservation of the momentum, which in the force-free limit reduces to the Poynting vector:

$$4\pi S_i = \epsilon_{ijk} E^j B^k. \tag{7.157}$$

After evolving the Poynting flux, the electric field can be reconstructed with the aid of the force-free condition (7.152), that is,

$$E_i = -\frac{4\pi}{B^2}\, \epsilon_{ijk} S^j B^k\,. \tag{7.158}$$

As in the case of the ideal MHD, neither the charge density q nor the gauge source Ψ appear on the equations, so they do not have to be evolved. Note also that in this limit we have neglected the effects of the fluid, so we do not need to evolve it at all, at least as far as we are just interested in the behavior of the electromagnetic field.

Finally, just mention that the characteristic speeds are given by two Alfven waves and two magnetosonic waves, moving at the speed of light. Thus, we can still use the expression (7.147) with $a = 1$.

7.5 Further developments

Current research in the field of fluids within general relativity is aimed at the study of a variety of different systems. Mostly these are systems containing strongly gravitating objects or related to cosmology, where gravity dominates on large scales. Among these we briefly review here some recent developments, focusing on the study of binary boson and neutron stars. Certainly the literature on these topics is vast, so we restrict ourselves to some representative examples where we have direct experience. We address the interested reader to more complete reviews, like for instance [18, 19] for boson stars and [13] for neutron stars.

7.5.1 Boson stars collisions

Boson stars have been studied in many different contexts [20]. They have been used as alternative models for compact objects, including black holes [21]. Another outstanding application is their use as models for dark matter [22].

The problem arises from the observed velocity curves of the stars in a galaxy, which cannot be explained by using only standard gravity and visible matter. Different alternatives to resolve this issue are being considered. Among them, there is the possibility of the presence of an additional 'dark' matter source, which can be modeled by boson stars, reproducing more accurately the observed velocity curves [23]. So, our main motivation in studying boson stars is to understand their possible role in explaining the dark matter content of the universe.

There are at least two interesting scenarios arising from the collision of the boson stars associated to galaxies. The first one would be to analyze the dynamics and interactions from the collision of two boson stars and compare with the observed collision of two galaxies, like for instance the bullet cluster [24]. On the other hand, it is commonly believed that most of the current galaxies have suffered some collision in their lifetimes, forming as a result larger and more stable galaxies. Consequently, the second issue would be to analyze the stability of the final objects produced by the collision of the boson stars.

Up to now, only the case where both boson stars are described with a single global scalar field ϕ has been considered. This choice fully determines the interaction between the stars, although other alternatives are being currently considered. The head-on collisions have been studied preliminarily in [25], and with more detail with an axial symmetry code in [6]. The results, even in this simple setup, showed very different behaviors depending on the masses and initial velocities of the stars. For instance, when the velocity of the boson stars at the collision is high enough, they cross each other keeping mainly the original shape after the interaction: a behavior which reminds us of a soliton.

On the other hand, when the velocities are low and the masses of the stars are large, the final object will be too massive to prevent the gravitational collapse and it will form a black hole. But when the masses are small, the dynamics and final object will be determined by the type of interaction. Let us analyze this issue in more detail following mainly the results described in [26]. The freedom introduced by the complex nature of the scalar field (i.e., the solution for the initial data is unaffected by either a phase shift δ or a frequency reflection $\epsilon = \pm 1$) allows many different initial configurations. For small masses and collision speeds, we can assume initial configurations for two boson stars of the form

$$\phi = \phi_0^{(1)}(\mathbf{r_1})\, e^{iwt} + \phi_0^{(2)}(\mathbf{r_2})\, e^{i(\epsilon w t + \delta)}. \qquad (7.159)$$

In the case of the collision of two identical boson stars (i.e., $\delta = 0$, $\epsilon = 1$) the simulations suggest that the merger results into a single boson star in the stable branch, which oscillates with large amplitude perturbations. The behavior of the collision of a boson star with another one in phase opposition (i.e., $\delta = \pi$, $\epsilon = 1$) resembles what one would expect for two objects subject to a repulsive force, but confined within a potential well: the objects oscillate around attractor positions. Elucidating the final fate of this system will require much longer and accurate evolutions. The last case considered was the collision of a boson star with an otherwise identical star except with the opposite Noether charge density (i.e., $\delta = 0$, $\epsilon = -1$). Such a star is called an anti-boson star and rotates in the opposite direction than its counterpart in the complex plane. Like in the phase opposition case, the early behavior agrees with that of the boson–boson case. As time goes on, however, notable differences arise. The dynamical behavior when the stars get close is weaker than in the boson–boson case but stronger than in the boson–phase opposition case.

The orbiting case has only been studied very recently in [27]. The cases considered did not show in the final state a stable stationary axisymmetric configuration.

7.5.2 Neutron stars collisions

Among the binary collisions of astrophysical bodies, the binary neutron star one is specially important for the production of both gamma ray bursts and strong gravitational waves. Analogous to the black hole case, the stars orbit for many orbits during the inspiral phase until they reach some critical distance, at which they merge. As opposed to the binary black hole case, the system now has further degrees of freedom, so the dynamics can be significantly different. For instance, at late stages in the inspiral phase, when the stars are close enough, tidal effects can produce deformations which lead eventually to matter transfer or even strong enough for disrupting one of the stars when the masses are very different.

Subsequently, the merger will produce a differentially rotating neutron star (DRNS), which can display very different behaviors depending on the mass, equation of state, and other effects like the presence of magnetic fields and the neutrino cooling. To date only the simplest issue, the mass dependence, has been analyzed in some detail [28]. For low-mass stars, the DRNS will be in the stable branch and it will evolve toward a uniformly rotating star (RNS) due either to viscous or some effective dissipative effects. The final RNS may end up on the unstable branch and collapse to a black hole. For moderate masses, the DRNS can have enough mass to become a black hole (i.e., being in the unstable branch), but the excess of angular momentum provides an effective extra pressure that prevents the collapse for some time. As the star loses angular momentum, it starts collapsing to a black hole. Finally, for high masses the DRNS collapses to a black hole soon after the merger [29]. Although the qualitative behavior is similar, the exact output and mass thresholds differ when using a more realistic EOS [30].

Numerical simulations of the merger of the orbiting binary NS, for very simple EOS and with low resolution, were performed before the binary BH ones. The absence of singularities, coupled to a weaker curvature scenario, seems to be the main reason. However, the lack of resolution prevented to completely trust the results, even in this simplistic scenario. The current use of advanced numerical techniques, like adaptive mesh refinement, allowed for much higher resolutions. The introduction of AMR in the GRHD codes opened a new line of more accurate evolutions with increasingly reliable extraction of the gravitational waves emitted during the merger, both for binary neutron star [31–33] and for mixed neutron star–black hole binaries [34]. Current efforts concentrate both on studying this problem with higher resolution and adding key missing physical ingredients, like the influence of the magnetic fields and the cooling due to neutrino transport.

The addition of magnetic fields, which in standard neutron stars are measured to be as large as 10^{12} Gauss, is necessary to attempt explaining some observed phenomena. The production of gamma ray bursts is nowadays associated with strong magnetic fields produced in compact objects, like the catastrophic events relating neutron stars and black holes. These large magnetic fields can have a significant effect on the dynamics of the star, which can be summarized into

- producing extra pressure and anisotropic effects;
- redistributing angular momentum, by connecting parts of the fluid which would be disconnected in the absence of magnetic fields.

In addition, the magnetic fields can be amplified during the merger of binary neutron stars due to different mechanisms. The simplest one is the winding of the magnetic field: when a star with a poloidal magnetic field is rotating differentially, part of the kinetic energy is converted into a toroidal magnetic field. Another very effective mechanism is driven by the Kelvin–Helmholtz instability. During the merger, there is a shear between the fluids of both stars. By the Kelvin–Helmholtz instability, this shear produces vorticity and this enlarges the magnetic fields. There are many new instabilities which were not present in the absence of magnetic fields, which can have a large influence on the binary neutron stars dynamics. During the merger, magnetic fields can increase until the point that their effects will be comparable to the hydrodynamical ones. This different dynamics will produce, consequently, a different gravitational wave emission [35, 36]. Up to now only few magnetized binary NS evolutions have been performed, and more work is needed in order to confirm and understand these results.

The above is just a illustration of how the addition of another extra physical process can affect the dynamics. Finally, let us just mention that there are still some missing ingredients which are basic to describe correctly all the processes involved in neutron stars dynamics. The heating of the plasma during the merger induces a large production of neutrinos, which carry away energy and angular momentum from that region of the star. The proper formalism for describing this radiation transport is given by the Boltzmann equations, which amount to solve a 7D system (four spacetime coordinates plus three momenta) of integro-differential equations. Since it is too expensive computationally, some approximation approaches have been suggested in order to deal with this problem. More theoretical work is needed in order to discern which approach is more convenient.

References

1. R. Ruffini and S. Bonazzola, Phys. Rev. **187**, 1767 (1969).
2. D. J. Kaup, Phys. Rev. **172**, 1331 (1968).
3. S. H. Hawley and M. W. Choptuik, Phys. Rev. **D62**, 104024 (2000).
4. G. Lemaitre, Rev. Mod. Phys. **21**, 357 (1949).
5. W. H. Press, B. P. Flannery, S. A. Teukolsky and W. T. Veterling, *Numerical Recipes*, Cambridge University Press, Cambridge (1989).
6. K. W. Lai, Ph. D. Thesis, University of British Columbia (2004), gr-qc/0410040.
7. M. Colpi, S. L. Shapiro and I. Wasserman, Phys. Rev. Lett. **57**, 2485 (1986).
8. K. O. Friedrichs, Commun. Math. Phys. **27**, 749–808 (1974).
9. A. M. Anile, *Relativistic Fluids and Magneto-Fluids: With Applications in Astrophysics and Plasma Physics*, Cambridge Monographs on Mathematical Physics (1989).
10. J. A. Font, J. A. Ibañez, J. M. Marti and A. Marquina, Astron. Astrophys. **282**, 304–314 (1994).
11. F. Banyuls et al., Astrophys. J. **476**, 221–231, (1997).
12. J. Read, B. D. Lackey, B. J. Owen and J. Friedman, ArXiv:0812.2163 (2008).
13. J. A. Font, Numerical hydrodynamics and magnetohydrodynamics in general relativity. Liv. Rev. Relat. **11** (2008). www.livingreviews.org/lrr-2008–7
14. C. Kittel, *Quantum Theory of Solids*, Wiley, New York (1987).
15. J. D. Bekenstein and E. Oron, Phys. Rev. **D18**, 1809 (1978).
16. G. Q. Chen, D. Levermore and T. P. Liu, Comm. Pure Appl. Math. **47**, 787–830 (1994).
17. T. Leismann, L. Aloy, M. A. Anton, E. Muller, J. M. Marti, J. A. Miralles and J. A. Ibañez, Astron. Astrophys. **436**, 503, (2005).
18. E. W. Mielke and F. E. Schunck, Class. Quantum. Grav. **20**, R301 (2003).
19. E. W. Mielke, B. Fuchs and F. E. Schunck, *Proceedings of the 10th Marcel Grossmann Meeting* (2006). astro-ph/0608526.
20. P. Jetzer, Phys. Rep. **220**, 163 (1992).
21. F. S. Guzman Phys. Rev. **D73**, 021501(R), (2006).
22. J. Lee and I. Koh, Phys. Rev. **D53**, 2236–2239 (1996).
23. F. E. Schunck and D. F. Torres, Int. J. Mod. Phys. **D9**, 601–618 (2000).
24. D. Clowe et al. Astrophys. J. **648**, L109–L113 (2006).
25. J. Balakrishna, Ph. D. Thesis. Washington University (1999). gr-qc/9906110.
26. C. Palenzuela, I. Olabarrieta, L. Lehner and S. Liebling, Phys. Rev. **D75**, 064005 (2007).
27. C. Palenzuela, L. Lehner and S. Liebling, Phys. Rev. **D77**, 044036 (2008).
28. M. Shibata and K. Uryu, Phys. Rev. **D61**, 064001 (2000).
29. S. L. Shapiro, Phys. Rev. **D58**, 103002, (1998).
30. M. Shibata, K. Taniguchi and K. Uryu, Phys. Rev. **D71**, 084021 (2005).
31. M. Anderson et al., Phys. Rev. **D77**, 024006 (2008).
32. T. Yamamoto, M. Shibata and K. Taniguchi, Phys. Rev. **D78**, 064054 (2008).
33. L. Baiotti, B. Giacomazzo and L. Rezzolla, Phys. Rev. **D78**, 084033 (2008).
34. Z. B. Etienne et al., Phys. Rev. **D77**, 084002 (2008).
35. M. Anderson et al., Phys. Rev. Lett. **100**, 191101 (2008).
36. L. Baiotti, B. Giacomazzo and L. Rezzolla, Phys. Rev. **D** (2009).

Index

Bona, C. et al.: *Index*. Lect. Notes Phys. **783**, 211–214 (2009)
DOI 10.1007/978-3-642-01164-1_BM2

GPSR Compliance

The European Union's (EU) General Product Safety Regulation (GPSR) is a set of rules that requires consumer products to be safe and our obligations to ensure this.

If you have any concerns about our products, you can contact us on ProductSafety@springernature.com

In case Publisher is established outside the EU, the EU authorized representative is:

Springer Nature Customer Service Center GmbH
Europaplatz 3
69115 Heidelberg, Germany

Batch number: 10682323

Printed by Printforce, the Netherlands